Springer Tracts in Mechanical Engineering

Springer Tracts in Mechanical Engineering (STME) publishes the latest developments in Mechanical Engineering - quickly, informally and with high quality. The intent is to cover all the main branches of mechanical engineering, both theoretical and applied, including:

- Engineering Design
- Machinery and Machine Elements
- Mechanical structures and stress analysis
- Automotive Engineering
- Engine Technology
- Aerospace Technology and Astronautics
- Nanotechnology and Microengineering
- Control, Robotics, Mechatronics
- MEMS
- Theoretical and Applied Mechanics
- Dynamical Systems, Control
- Fluids mechanics
- Engineering Thermodynamics, Heat and Mass Transfer
- Manufacturing
- Precision engineering, Instrumentation, Measurement
- Materials Engineering
- Tribology and surface technology

Within the scopes of the series are monographs, professional books or graduate textbooks, edited volumes as well as outstanding PhD theses and books purposely devoted to support education in mechanical engineering at graduate and post-graduate levels.

More information about this series at http://www.springer.com/series/11693

Sine Leergaard Wiggers · Pauli Pedersen

Structural Stability and Vibration

An Integrated Introduction by Analytical and Numerical Methods

 Springer

Sine Leergaard Wiggers
Department of Technology and Innovation
University of Southern Denmark
Odense M
Denmark

Pauli Pedersen
Department of Mechanical Engineering
Technical University of Denmark
Kgs. Lyngby
Denmark

ISSN 2195-9862　　　　　　ISSN 2195-9870　(electronic)
Springer Tracts in Mechanical Engineering
ISBN 978-3-319-89202-3　　　　ISBN 978-3-319-72721-9　(eBook)
https://doi.org/10.1007/978-3-319-72721-9

To Hanne, Søren, Sara, and Sofia

Preface

This book is based on lecture notes for a postgraduate course at University of Southern Denmark (SDU). For us, the integrated teaching of stability and vibration justifies the efforts needed to join many different subjects in a presentation on the introductory level.

Basic understanding of the content of the course, named Stability and Vibration, can be obtained from the derived formulas and graphical presentations of numerical results that give an overview of parameter dependence. Students may reproduce these graphs with their personal MATLAB programs and obtain further results and a deeper insight. Formulas for instability modes and vibrational modes should be used to write interactive and dynamic programs to get a deeper insight into the subject of the course.

The chosen presentation to a large extent in non-dimensional quantities may at first seem disturbing, but it gives more generality and is hopefully appreciated after some time. It should be added that dimensional check of formulas is still possible.

A number of different notes written in Danish are behind this book. Critics of the present version and suggestions to improve the book are most welcome on email to slp@iti.sdu.dk.

Denmark
September 2017

Sine Leergaard Wiggers
Pauli Pedersen

Contents

1 Introduction 1

2 Beam-Column Differential Equation 5
    2.1   What Is a Beam-Column 5
    2.2   Differential Equation for Static Equilibrium 7
    2.3   Differential Equation for Eigenfrequencies 8
    2.4   Names and Symbols for Boundary Conditions (BC) 10

3 Eigen Solutions for the Euler Cases 13
    3.1   Boundary Conditions 13
    3.2   How to Solve an Eigenvalue Problem 14
    3.3   Instability Modes for the Euler Cases 14
            3.3.1   Instability Mode for the Euler Case I 14
            3.3.2   Instability Modes for the Euler Cases II–IV ... 15
            3.3.3   Instability Mode for the Euler Case V 16
    3.4   Eigenfrequency Modes for the Euler Cases 17
            3.4.1   Eigenfrequency Modes for the Euler Case I 17
            3.4.2   Eigenfrequency Modes for the Euler Cases II–VI ... 18
    3.5   Summary 21

4 Beam-Columns and Applied Berry Functions 23
    4.1   Model of Beam-Column 23
    4.2   General Moment Loads 26
    4.3   Elastic Support Against End Rotations 28
            4.3.1   A Fixed Support as a Limiting Case 30

5 Shear Beam Loads and Cantilever Beam-Columns 33
    5.1   Shear Loads on Beam-Columns 33
            5.1.1   A Fixed Support as a Limiting Case 35
    5.2   Cantilever Beam-Columns 36
            5.2.1   Two Cantilever Cases 38

6 Beam-Column Eigenfrequencies 41
  6.1 Mathmatical Model for Different Physical Problems 41
  6.2 Solution of DE (6.1) with BC (6.2) 42
  6.3 Eigenfrequency as a Function of Conservative Axial Load 44
  6.4 Rotational Spring Supports 45
     6.4.1 Euler Cases as Limiting Cases 46

7 Buckling with Spring Supported BC 47
  7.1 Mathematical Definition and Physical Experiments 47
  7.2 Different Instability Formulations 48
  7.3 Buckling with End Rotations 49
  7.4 Buckling with End Translations 51
  7.5 Buckling with Winkler Support 53
     7.5.1 Eigenfrequencies with Winkler Support 57

8 Eigenfrequencies of Beam-Columns with Spring Supported BC 59
  8.1 Eigenvalue Problems with Analytical Solution 59
  8.2 Solving the Transcendental Equations 62
  8.3 Explicit Solutions by Inverse Approach 63
     8.3.1 Lowest Eigenfrequency as a Function of Support Stiffnesses, Assumed No Column Force 64
  8.4 Alternative Function Expressions 64
  8.5 Specific Graphically Presented Results, Obtained by the Newton–Raphson Method 66
     8.5.1 Lowest Eigenfrequency as a Function of Column Force, with Support Stiffnesses as Parameters 68
     8.5.2 Lowest Eigenfrequency as a Function of Column Force, Further BC Parameters 68
  8.6 Lowest Eigenfrequency as a Function of Non-conservative "follower" Column Force 70

9 Dynamic Stability Formulation 73
  9.1 One and Two Degrees of Freedom 73
  9.2 An Elementary Beam-Column Case 74
  9.3 Column with a Point Mass 77
  9.4 An Improved Dynamic Column Model 80
  9.5 Non-conservative Column Load 82

10 Stability of 2D Frames 85
  10.1 Frames 85
  10.2 Only One Beam-Column in the Frame 85
  10.3 Several Beam-Columns in the Frame 87
  10.4 Solution Procedure ("cookbook") 89
  10.5 Post-Critical Imperfection Analysis 91

**11   Buckling Stresses, Material Nonlinearity,
and Beam Modeling**   95
   11.1   Various Concepts   95
   11.2   Maximum Stress in Beam-Columns   96
   11.3   Material Nonlinearity   99
          11.3.1   Different Suggested Formulas   99
          11.3.2   Experiments and Safety Factor 100
          11.3.3   Design Formulas Used in Truss Optimization 101
   11.4   Improved Beam Modeling. 102
          11.4.1   Influence on Column Buckling 103
          11.4.2   Modeling Conclusions 104

**12   Large Displacements, Pre-buckling Strains,
and Snap-Through** 105
   12.1   Large Displacements for the Elastica 105
   12.2   Pre-buckling Strain and Modified Data. 108
          12.2.1   A Build Up Column. 109
          12.2.2   Notes on Global and Local Buckling
                   and Vibrational Modes 111
   12.3   Linear Elastic Snap-Through 112

13   Dynamics of Discretized Linear Systems 115
   13.1   Discretization Continuous to Discrete. 115
   13.2   Dynamic Equilibrium 116
   13.3   Non-damped, Free Vibrations 117
          13.3.1   Real, Positive Eigenfrequencies. 117
          13.3.2   Orthogonal Eigenmodes 118
   13.4   Expansion for More General Linear Systems 119
   13.5   Vibrations with Single dof. 120
          13.5.1   Non-damped Single dof Vibrations 120
          13.5.2   Damped Single dof Vibrations 121

14   Discretized Stability Analysis 123
   14.1   Different Stiffness Matrices 123
   14.2   Green-Lagrange Strain Model 124
          14.2.1   Finite Element Equilibrium and the Secant
                   Stiffness Matrix 125
          14.2.2   Residuals and Newton–Raphson Iterations 126
          14.2.3   The Stress Stiffness Matrix 126
          14.2.4   The Tangent Stiffness Matrix for a Tetrahedron
                   Element. 127
   14.3   Linear Bifurcation Buckling 128
          14.3.1   Nonlinear Obtained Reference State 129

14.4 Nonlinear Implicit Procedure or Nonlinear Explicit
      Procedure 129
14.5 Dilatation Errors from Linear Strain Modeling 131

15 Routh–Hurwitz-Liénard–Chipart Criteria 133
15.1 Relation to Dynamic Stability 133
15.2 Necessary Criterion 134
15.3 Necessary and Sufficient Criteria 135
15.4 Criteria Expressed by Polynomial Coefficients 136
15.5 Polynomials with Complex Coefficients 138
15.6 Example with Complex Coefficients 139

16 Numerical Tools 141
16.1 Newton–Raphson Iterations for Solving Nonlinear
      Problems 141
16.2 Numerical Tools for Linear Problems 142
16.3 Gauss Factorization 142
16.4 Linear Solutions by Forward and Backward
      Substitutions 144
16.5 Superelement Technique 145
16.6 Power Method and Inverse Iteration 146
      16.6.1 Rate of Convergence 148
      16.6.2 Orthogonalization and Shift 149
16.7 Subspace Iteration 151

References 153

Index 155

Acronyms

Traditional notations are normally preferred. For non-dimensional quantities, the use of the Greek letters is preferred, but with exceptions as for non-dimensional boundary stiffnesses. For dimensional quantities, the use of the Latin letters is preferred. Matrix notation is used for linear algebra with [ ] for rectangular and quadratic matrices, {} for column matrices and $\{\}^T$ for row vectors, i.e., T as upper index for transpose.

Notations

Latin notations, mainly dimensional quantities

A	Cross-sectional area
A_{min}	Minimum allowable cross-sectional area
$B, \bar{B}, \tilde{B}$	Berry functions
C	Lower index for critical quantity
C, C_1, C_2, C_3, C_4	Constants in a solution
c	Cross-sectional position for largest axial stress in (11.5)
D	Determinant and characteristic function
$D_1, D_2, D_3, D_4, D_5, D_6, D_7$	Sub-functions of D
D, d	Diameters of a circular cross section
E, E_t, E_s, E_r	Moduli of elasticity, specifically, Young's, tangent, secant, and reduced
e	Eccentricity length
F	Function for differentiation of finite integral in (2.4)
$f, f_{i,j}$	Flexibility components

G	Shear modulus of elasticity
g	Acceleration of gravity
$g, \bar{g}$	Parameters defined by (8.12)
h	Height
I, i	Cross-sectional moments of inertia
i	Imaginary unit $= \sqrt{-1}$
K	Stiffness for translational spring at boundary
k, k_C	Non-dimensional K and critical value of this
L, l	Lengths of a beam
$\tilde{L}$	Deformed length of a bar
l_e	Effective length in buckling or in vibration
M	Moment, internal or external load
M_0, M_1	Moments at the boundary of a beam-column
M_A, M_{AB}	Moment at a point A specifically of beam-column between A, B
m	Distributed moment per length
N	Axial column force, positive in tension
N_C	critical axial column force
N_{AB}	Axial column force in beam-column between A, B
n	Number, say 0, 1, 2,...
n_C	Number of buckles in critical state
P, P_C	Axial column compressive force, and critical value of this
P_p	Compressive force resulting in compressive stress σ_p
p	Volume force
$P(z), \tilde{P}(z)$	Complex polynomial and its conjugated, see (15.18)
$Q(z)$	Real coefficient part of a complex polynomial
$R(z)$	Imaginary coefficient part of a complex polynomial
Q	Beam external force in transverse direction
q	Distributed beam external force in transverse direction per length
r	Stiffness per length
S_0, S_1	Stiffness for rotational springs at boundary ends 0, 1
s_0, s_1	Non-dimensional S_0, S_1
$s, s_{i,j}$	Stiffness components
$(s_0)_C$	Critical non-dimensional stiffness at 0
s	Length position at a curved beam-column
T	Internal shear force for a beam-column
$T_0.T_1$	Shear forces at boundary ends 0 and 1
t	Time parameter
x	Position parameter in the length direction of a beam-column

y	Displacement in the transverse direction, see Fig. 2.1, i.e., $y = y(x)$
z	Polynomial complex parameter

Greek notations, mainly non-dimensional quantities

α	Stability parameter in the exponential time function $e^{(\alpha + i\omega)t}$
α	Factor in trigonometric functions
α	Value of angle in the Elastica
α	Parameter defined by (6.7)
β	Parameter defined by (6.7)
Γ	Slenderness ratio
γ	Parameter for non-conservative load
γ	Shear strain
Δ	Small but finite incrementation prefix
δ	Virtual prefix
δ	Displacement at a specific point
∂	Partial prefix
$\varepsilon_{i,j}$	Strain component
$\varepsilon, \varepsilon_C$	Axial strain and critical value of this
ζ	Factor in (9.21) for mass moment of inertia
η	Non-dimensional position parameter for load
$\eta_{i,j}$	Green-Lagrange strain component
θ	Angle of cross-sectional rotation (radians positive anticlockwise)
θ_0, θ_1	Angles for end rotation at 0 and 1
κ	Curvature of a beam
Λ	Modified eigenvalue in (11.25)
$\lambda, \bar{\lambda}, \lambda_C$	Non-dimensional column loads and critical value
μ	Cross-sectional parameter for maximum shear stress
μ_E	Factor of safety in the Euler range
ν	Poisson's ratio
$\xi, \bar{\xi}$	Non-dimensional position parameter $\xi = x/L$ and specifically integration parameter
ρ	Density parameter
ρ	Non-dimensional inertia radius
$\sigma_{i,j}$	Stress component
σ, σ_C	Axial stress and critical value of this
σ_p	Compressive stress at proportionality limit
σ_y	Compressive yield stress
τ	Shear stress
ϕ, ϕ_C	Non-dimensional squared eigenfrequency and its critical value

ϕ Non-dimensional for magnetic attraction or for Winkler support
ϕ, ψ Angles in frame examples
ω^2, ω_C^2 Squared eigenfrequency and its critical value

Matrix notations, mainly for numerical formulations

$\{A\}, \{\tilde{A}\}$ Static and dynamic load vector
$[B_t]$ Strain–displacement relation
$[C]$ Damping matrix, with damping coefficients $c_{i,j}$
$\{D\}, \{\tilde{D}\}$ Static and dynamic displacement vector
$\{\Delta_r\}$ Eigen vector for mode r
$[F]$ Flexibility matrix, with flexibility coefficients $f_{i,j}$
$[L_s]$ Constitutive secant matrix, with coefficients $(l_{i,j}), s$
$[L_t]$ Constitutive tangent matrix, with coefficients $(l_{i,j}), t$
$[M]$ Mass matrix, with mass coefficients $m_{i,j}$
$\{0\}$ Null vector, where all component are zero
$\{R\}$ Residual vector, for Newton–Raphson iterations
$[S]$ Stiffness matrix, with stiffness coefficients $s_{i,j}$
$[S_0]$ Initial stiffness matrix for linear elasticity
$[S_s]$ Secant stiffness matrix
$[S_\gamma]$ Displacement gradient stiffness matrix
$[S_\sigma]$ Stress stiffness matrix
$[S_t]$ Tangent stiffness matrix $[S_\gamma] + [S_\sigma]$

Special symbols and abbreviations

BC   Boundary Condition(s)
DE   Differential Equation(s)
dof   Degree of freedom
FE   Finite Element (method)
d Prefix for differential
$\prime$ Partial differentiation with respect to non-dimensional position coordinate
$\cdot$ Partial differentiation with respect to time
$\|\ \|$ Determinant or norm value
$:=$ By definition
$\sim$ Indicate dynamic dependence or alternative quantity

Chapter 1
Introduction

This small book covers the subjects of stability, of vibration, and of dynamic stability, in the first chapters restricted to 2D beam problems. The book is written for a course named stability and vibration, and the intention is to communicate the similarities and interactions of the three subjects that are more traditionally treated separately.

Another non-traditional aspect is the focus on spring supports with graphical results to clearly understand the importance of support modeling. Initial analysis of continuous models is performed by a solution of differential equations, but with finite element (FE) analysis, the generality is rather unlimited.

Why this book?

A larger number of books are written on stability, and an even larger number are written on vibration. References to such books are listed in the list of references at the end of the present book. The two practical important subjects of stability and vibration are seldom treated as an integrated subject, although closely related. The book by Ziegler (1968) is well suited for teaching advanced subjects and contains a clear classification of our different physical problems. The book by Panovko and Gubanova (1964) supports the good tradition of studying very simplified models of physical problems, in order to focus on the behavior in question. A recent book on advanced vibrations and stability theory is by Thomsen (2003). This book focuses on nonlinear theory with chaos theory and special high-frequency effects.

Equal focus on stability and vibration

To integrate stability and vibration on the introductory level is a primary goal of the present book. A second goal of the book is to focus also on non-classical boundary cases for beam problems, i.e., to extend the cases termed Euler cases. Linear elastic rotational springs as well as translational springs are modeled from an analytical point of view by extended use of well-known functions, and the support spring stiffnesses are treated as parameters. From this analysis, a number of functional relations are obtained:

© Springer International Publishing AG, part of Springer Nature 2018
S. L. Wiggers and P. Pedersen, *Structural Stability and Vibration*, Springer Tracts
in Mechanical Engineering, https://doi.org/10.1007/978-3-319-72721-9_1

Extension to linear elastic supports

- Stability as a function of boundary conditions (BC)
- Eigenfrequency as a function of BC.
- Eigenfrequency as a function of column force.

Graphical result display

The intention is to graphical display such results, often obtained by explicit expressions. The book is not an alternative to the many written books, but should be seen as a supplement to these.

It is the intention, generally to let these graphical displays dominate, not for reading specific numerical values from the graph but to illustrate the principal relations between stability (column force), eigenfrequency (with focus on the lowest one), and the parameters of the boundary conditions (BC). Computer tools make this possible and influence the final look of the book, hopefully liked by the readers.

Chosen notations and sign

Traditional notations are normally preferred, as seen in the list of symbols. For non-dimensional quantities, the use of the Greek letters is preferred, but with exceptions as for non-dimensional boundary stiffnesses. For dimensional quantities, the use of the Latin letters is preferred. Matrix notation is used for linear algebra with [ ] for rectangular and quadratic matrices, { } for column matrices and $\{\}^T$ for row vectors, i.e., T as upper index for transpose.

Sign decision is in the direction of the axes of a Cartesian coordinate system, for moments (rotations) as well as for forces (displacements). The book is for the continuous models mainly limited to 2D problems and thus moment (rotation) is chosen anticlockwise.

2D-beam-columns

The initial part of the book is analytical oriented with a focus on beam-columns. Later, the general formulation is related to finite element (FE) models, and the structural or continuum models are then generalized.

Finite element models

A structure/continuum may be described by system matrices like mass matrix $[M]$, stiffness matrix $[S]$, and stress stiffness matrix $[S_\sigma]$. Then, in addition to analysis also synthesis aspects can be involved, such as design for vibration and stability. This includes sensitivity analysis that is determination of response gradients as a function of design parameters.

Design for stability/vibration

Additional insight into the subject of stability as well as that of vibration is hopefully obtained by the integrated treatment. It is the intention that each formula should to a large extent be derived to convince the students and other readers.

Book layout

The layout of the book shows the wish to jointly treat stability and vibration, here on the introductory level and with simple beam theory as the continuous structural model. In Chap. 2, the intention is therefore to derive the fourth order differential equation (DE) where the influence from column force and eigen vibration is included.

Euler cases, Eigenvalue problems, and Characteristic equations

In Chap. 3, the homogeneous part of the DE with specific BC gives classical important eigenvalue results on stability (critical compressive column forces) as well as on vibration (eigenfrequencies without column force). Six cases are known as Euler cases, and their solution involves eigenvalues and corresponding eigenfunctions, i.e., critical instability buckling modes and vibration modes for beam bending. Fourth-order (max) determinants are involved to obtain characteristic equations from which the eigenvalues are determined. All the results of this mathematical exercise are presented in one page of three tables.

Solutions to non-homogeneous versions of the DE with BC are initiated in Chap. 4 by the beam-column theory, i.e., beam theory which include bending moments from a column force by setting up static equilibrium in a displaced state. This problem is still linear in the bending loads but nonlinear in the column force. The solution to these problems is described by a number of trigonometric/hyperbolic functions, in the present book commonly named Berry functions. Rotational end springs are introduced and thus extension of BC from classical simply, fixed and free BC.

Nonlinear beam-column models

Shear beam loads and cantilever beam-columns are introduced in Chap. 5, which also include the solution to the most simple case of eigenfrequency as a function of column load ($\omega^2 = \omega^2(\lambda)$). Taking translational end springs into account complicates the analysis and also changes the principal character of the resulting functional relations.

In Chap. 6, the beam-column eigenfrequencies are graphically presented for different combinations of spring support.

Buckling for spring supports

Chapter 7 treats beam buckling with spring supported BC. It is initiated with a general discussion of the concept of stability, comparing results of mathematical analysis with observed physical experiments. Winkler support is the name for continuous translational support, and the chapter includes the influence of this support on buckling as well as on eigenfrequencies. Of specific interest for this case is that the lowest number of critical buckles may be larger than one.

Frequency for spring supports

It may be stated that Chap. 8 includes most of the preceding chapters as special cases, because it generalizes to eigenfrequencies with spring supported BC and analyzes eigenfrequencies as a function of column force. A general characteristic function is formulated with three support stiffness parameters combined with seven sub-functions that only depend on non-dimensional column force and non-dimensional squared eigenfrequency. Results for a number (12) of principal different BC are presented, and specific results are graphically displaced. The chapter is restricted to conservative problems, except for a final case.

Non-conservative dynamic stability

Non-conservative problems are introduced in Chap. 9, which is written as an introduction to dynamic stability. The essential difference between divergence instability with zero eigenfrequency and flutter instability with vibration of increasing amplitude is clarified and exemplified by a most simple cantilever beam problem with a tip mass.

Methods for 2D-frames

Returning to only divergence instability, Chap. 10 discusses analytical methods for determination of critical forces for rather simple 2D frames. A final example analyzes the Roorda–Koiter frame to exemplify the notion of imperfection sensitive structures.

Imperfection sensitivity, Material, and geometrical nonlinearity

Chapter 11 relate to buckling stresses and material nonlinearity, while Chap. 12 relate to geometrical nonlinearity.

Discretized models

With discretized models such as FE models, general models can be analyzed for stability as well as for vibrations. In Chap. 13 on dynamics of discretized models and Chap. 14 for discretized stability analysis, the matrix formulation is dominating.

Theoretical and numerically tools

Chapter 15 is a chapter with the theory behind the important stability criteria, named Routh–Hurwitz-Lienard–Chipart criteria, is added.

Chapter 16 is a chapter on numerical tools.

Chapter 2
Beam-Column Differential Equation

After classifying the modeling names of bar, column, beam, and beam-column, the important differential equations with boundary conditions (DE with BC) are derived. For this mathematical formulation of physical problems, the knowledge to set up static equilibrium is essential. It is important that the reader takes the time to be familiar with the results and the basic assumptions behind the theory. Finally, the chosen names and symbols for the treated BC are stated. Dimensionless critical column load λ is defined, being negative as the axial force N is defined positive for a tensile force. Note that for beam-columns tensile forces are also of interest. Dimensionless eigenfrequency ω is defined by its squared value.

2.1 What Is a Beam-Column

Beam-column is the name for a specific model within the subject of solid mechanics. Like most models, the concept of beam-column includes the description of the structural element as well as its use, i.e., the assumed load case(s). To illustrate this model concept, the definition of other close models is

- Bar is "a long continuum", loaded with pure tension
- Column is "a long continuum", loaded with pure compression
- Beam is "a long continuum", loaded with bending loads
- Beam-column is "a long continuum", loaded with tension or compression together with bending loads

The simple column theory is based on the differential equation for Bernoulli–Euler beams, also termed simple beam theory

$$M(x) = EI(x)\frac{d^2 y}{dx^2} \tag{2.1}$$

© Springer International Publishing AG, part of Springer Nature 2018
S. L. Wiggers and P. Pedersen, *Structural Stability and Vibration*, Springer Tracts
in Mechanical Engineering, https://doi.org/10.1007/978-3-319-72721-9_2

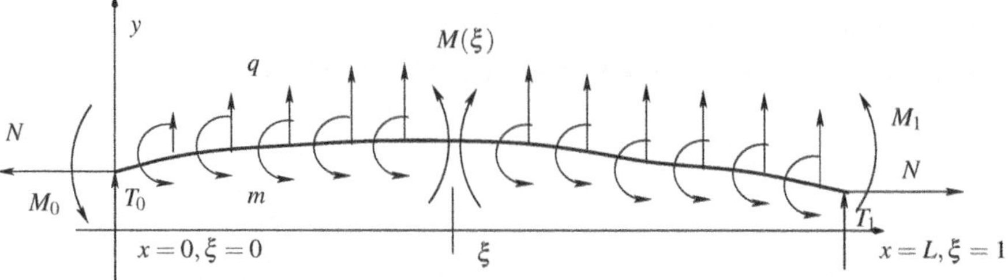

Fig. 2.1 Model of a beam-column for deriving equations for displaced equilibrium

where $M(x)$ is the internal moment at position x, E is modulus of elasticity, $I(x)$ is the cross-sectional moment of inertia at position x, and y is the displacement in the y-direction, all shown in Fig. 2.1, where the external beam forces are end moments M_0, M_1, end transverse forces T_0, T_1, distributed moments per length m (exemplified by rotational inertia) and distributed forces per length q. Furthermore, the axial force N is the external column force. Note that only for small displacements are forces in the y-direction identical to shear forces and forces in the x-direction are longitudinal forces.

Assumptions

The model result (2.1) is properly one of the most important equations in the subject structural mechanics, so the basic assumptions behind this model should be stated

- Plane cross sections remains plane and perpendicular to the beam axis
- Linear elasticity, i.e., strain ε proportional to stress σ
- Curvature approximated with d^2y/dx^2
- Small displacements, that with cross-sectional rotation $\theta \ll 1$
  gives $\sin\theta \simeq \tan\theta \simeq \theta$ and $\cos\theta \simeq 1$
- Shear strains from shear loads are neglected
- Strains in axial direction directly from the axial loads (compression or tension) are neglected

Displacement equilibrium

In the theory for column, the static equilibrium is formulated for the displaced structure (column). This implies that the loads originally in the direction of the beam axis (compression or tension) add to the bending loads.

Figure 2.1 shows a beam-column and a Cartesian coordinate system (x, y) or (ξ, y), as a non-dimensional length parameter (position parameter) $\xi = x/L$ is defined. The notion y' is used for $\partial y/\partial\xi$ and $y, y', y'', y''', \ldots$ all have dimension of length, and the equilibrium in the displaced state makes the solution nonlinear.

$$y' = \partial y/\partial\xi \;\Rightarrow\; dy/dx = y'/L, \;\; d^2y/dx^2 = y''/L^2, \;\; d^3y/dx^3 = y'''/L^3, \ldots$$

$$(2.2)$$

2.2 Differential Equation for Static Equilibrium

Static equilibrium for both left and right part structure, relative to the separation point at ξ can directly be seen from Fig. 2.1 and with (2.1), the result is

$$M(\xi) = EIy''/L^2$$

$$= -M_0 + T_0 L\xi + N(y(\xi) - y(0)) + L^2 \int_0^\xi q(\bar{\xi})(\xi - \bar{\xi})d\bar{\xi} - L \int_0^\xi m(\bar{\xi})d\bar{\xi}$$

$$= M_1 + T_1 L(1 - \xi) + N(y(\xi) - y(1)) + L^2 \int_\xi^1 q(\bar{\xi})(\bar{\xi} - \xi)d\bar{\xi} - L \int_\xi^1 m(\bar{\xi})d\bar{\xi} \tag{2.3}$$

where $\bar{\xi}$ is the integration variable, for the left part structure $0 \le \bar{\xi} \le \xi$ and for the right part structure $\xi \le \bar{\xi} \le 1$.

Using the formula for differentiation of a finite integral

$$\frac{d}{d\xi} \int_{g(\xi)}^{h(\xi)} F(\xi, \bar{\xi})d\bar{\xi} = \int_{g(\xi)}^{h(\xi)} \frac{\partial}{\partial \xi} F(\xi, \bar{\xi})d\bar{\xi}$$

$$+ \frac{dh(\xi)}{d\xi} F\left(\xi, \bar{\xi} = h(\xi)\right) - \frac{dg(\xi)}{d\xi} F\left(\xi, \bar{\xi} = g(\xi)\right) \tag{2.4}$$

in differentiating (2.3) with respect to ξ, still with $\bar{\xi}$ as integration variable, gives

$$(EIy'')'/L^2 = T_0 L + Ny' + L^2 \int_0^\xi q(\bar{\xi})d\bar{\xi} - Lm(\xi)$$

$$= -T_1 L + Ny' + L^2 \int_\xi^1 -q(\bar{\xi})d\bar{\xi} - Lm(\xi) \tag{2.5}$$

Performing further differentiations on (2.5) with respect to ξ the resulting differential equation is independent of the external end loads M_0, M_1 and T_0, T_1.

$$(EIy'')'' = NL^2 y'' + L^4 q - Lm' \tag{2.6}$$

Non-dimensional axial force

For constant bending stiffness EI (uniform beam), the axial force N is described non-dimensional by λ, with tension as positive like the normal decision for the axial force N (not only compressive forces are here of interest)

$$\lambda := \frac{NL^2}{EI} \tag{2.7}$$

Uniform beam

For constant, EI (2.5) is then written

$$y'''' - \lambda y'' \equiv \frac{L^4 q}{EI} - \frac{L^3 m'}{EI} \tag{2.8}$$

The statical boundary conditions for EI constant follows from (2.3) and (2.5)

$$y''(0) = \frac{-M_0 L^2}{EI}, \quad y''(1) = \frac{M_1 L^2}{EI},$$

$$y'''(0) - \lambda y'(0) = \frac{T_0 L^3}{EI} - \frac{L^3 m(0)}{EI}, \quad y'''(1) - \lambda y'(1) = -\frac{T_1 L^3}{EI} - \frac{L^3 m(1)}{EI} \tag{2.9}$$

Static boundary conditions

In test for correct dimensions of this section, it should be remembered that y, y', y'', y''', y'''' all have dimension of length.

2.3   Differential Equation for Eigenfrequencies

A dynamic problem may be converted to a static problem by the d'Alembert principle. The inertia forces are equivalent with volume forces p, that with an assumption of periodic vibrations with a separated exponential time function $y(x, t) = y(x)e^{i\omega t}$ (frequency ω) gives $\partial^2 y(x, t)/\partial t^2 = -\omega^2 y(x)e^{i\omega t}$ and by d'Alembert principle the equivalent static volume force (force per volume)

$$p = -\rho \frac{\partial^2 y}{\partial t^2} = \omega^2 \rho y \tag{2.10}$$

where ρ is mass density, t is time and y displacement amplitude in the y-direction. For the beam and beam-column, this corresponds to a force per length q

$$q = \omega^2 \rho A y \tag{2.11}$$

with A as the cross-sectional area.

Mathematical eigenvalue problem

For a simple model assume that the distributed moments m are zero and EI constant, then inserting (2.11) in (2.8) gives the differential equation

$$y'''' - \lambda y'' - \omega^2 \frac{L^4 \rho A}{EI} y \equiv 0 \quad \rightarrow$$

$$y'''' - \lambda y'' - \phi y \equiv 0 \tag{2.12}$$

defining the non-dimensional frequency parameter ϕ by

$$\phi := \frac{\omega^2 L^4 \rho A}{EI} \tag{2.13}$$

The differential equation (DE) (2.12) with boundary conditions (BC) may be seen as an eigenvalue problem with the pair λ, ϕ as an eigenvalue. The solutions then give eigenfrequencies ω as a function of axial loads or alternatively the stability load as a function of vibrating frequency. Note that zero vibrating frequency identifies the critical load.

Simply supported

The resulting differential equation (2.12) will first be solved for simply supported boundary conditions

$$y(0) = y(1) = y''(0) = y''(1) = 0 \tag{2.14}$$

Other eigenvalue problems than (2.12) with (2.14) will be discussed and solutions shown. The initial goal is to show the solutions for classical boundary conditions other than (2.14), with focus on the pure stability problem $\phi = 0$ and on the pure vibration problem $\lambda = 0$. The very practical problems of elastic supports are treated in Chaps. 7 and 8. More complicated cases are treated as specific topics, that relives many assumptions of this introductory chapter.

Different physical problems

Solutions to different eigenvalue problems of interest for the present book are

- 1. Static instability (divergence) by $\lambda \neq 0$ but $\phi = 0$
- 2. Eigenfrequency vibration by $\phi \neq 0$ but $\lambda = 0$
- 3. Eigenfrequency as a function of axial load $\phi \neq 0$ and $\lambda \neq 0$
- 4. Dynamic instability (flutter) by $\lambda \neq 0$ and $\phi \neq 0$

Simple cases of group 1 and 2 are treated in Chap. 3, group 3 in Chap. 8, and group 4 in Chap. 9.

2.4   Names and Symbols for Boundary Conditions (BC)

Names and symbols for boundary conditions (BC) are different in different books.

The present section will therefore list the chosen names and symbols and simultaneously show the many different BC that are studied analytically with respect to stability and vibration.

Figure 2.2 shows 12 different BC for beam-columns, later treated analytically. Of these cases, five different BC are normally named Euler BC, that are analyzed in Chap. 3. The more general comments to Fig. 2.2 are

- The name simply is chosen and is identical to the names pinned and hinged.

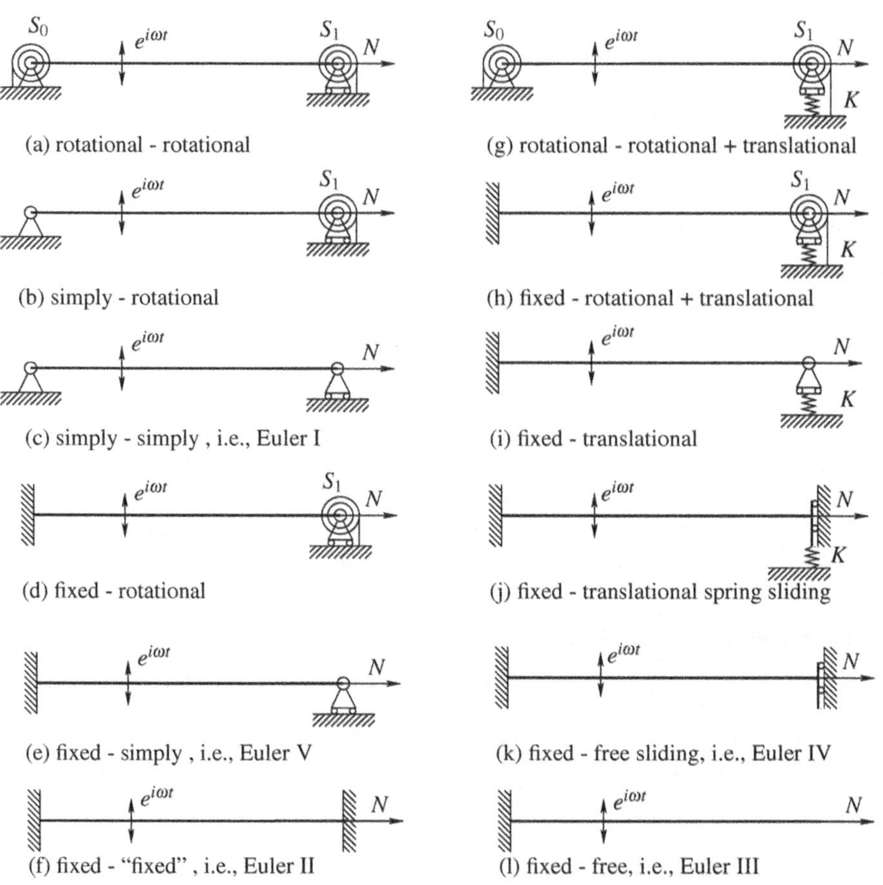

(a) rotational - rotational

(b) simply - rotational

(c) simply - simply , i.e., Euler I

(d) fixed - rotational

(e) fixed - simply , i.e., Euler V

(f) fixed - "fixed" , i.e., Euler II

(g) rotational - rotational + translational

(h) fixed - rotational + translational

(i) fixed - translational

(j) fixed - translational spring sliding

(k) fixed - free sliding, i.e., Euler IV

(l) fixed - free, i.e., Euler III

Fig. 2.2 Names and symbols for boundary conditions (BC) of the studied beam-column cases. Uniform beam-columns with constant EI and length L, non-dimensional length by $\xi = x/L$

- The left end of the beam-column is at the origin of the applied xy-coordinate system. The label of this point is 0, and this end point is not movable in either x or y direction.
- The label of the other end point of 1 as the non-dimensional coordinate ξ at this end is 1. An external column force (axial force) N is subjected to this point, and for all BC, the beam-column is free to move in the direction of N, although the symbols at this end may be stated as fixed, say in Euler cases II and IV.
- Elastic BC has an important influence on resulting eigenvalues for stability as well as for squared eigenfrequencies. Therefore, the efforts in analyzing the dependence on rotational end stiffnesses and the translational end stiffness, here chosen at the right end.
- The group of cases to the left in Fig. 2.2 all have no translational possibilities at the ends, i.e., $K = \infty$. The study of these 6 cases includes possible combinations of rotational stiffnesses S_0, S_1.
- The group of cases to the right in Fig. 2.2 have the translational possibility at the right end, and as seen five cases have fixed left end. Especially, translational displacement of the point of external column force has a strong influence on stability and vibration. The fact that the eigenvalue analysis is more complicated for these cases is the background for Chaps. 7 and 8 in addition to the Euler cases in Chap. 3.

Chapter 3
Eigen Solutions for the Euler Cases

Specific cases with constant bending stiffness EI and non-elastic BC are named the Euler cases, and results for these are included in most textbooks on stability and vibration. Instability modes and eigenfrequency modes for bending of the Euler cases are presented in tables with corresponding critical instability load and corresponding eigenfrequencies for the first five eigenfrequencies.

3.1 Boundary Conditions

The Euler cases refer to simple non-elastic boundary conditions (BC) and constant bending stiffness $EI \neq 0$, i.e., the cases c, f, k, i, and e in Fig. 2.2.

The kinematically BC are for non-translational end conditions no displacement $y = 0$ and for non-slope end conditions no gradient $y' = 0$. The statically BC are taken from (2.9) for no distributed external moments $m \equiv 0$. Thus without end moments $M = 0 \rightarrow y'' = 0$ and without end forces $T = 0 \rightarrow y''' = 0$. For the treated six Euler cases, the combined BC are:

I: simply–simply by $y(0) = y''(0) = y(1) = y''(1) = 0$

II: fixed–"fixed" by $y(0) = y'(0) = y(1) = y'(1) = 0$

III: fixed–free by $y(0) = y'(0) = y''(1) = y'''(1) - \lambda y'(1) = 0$

IV: fixed–free sliding by $y(0) = y'(0) = y'(1) = y'''(1) = 0\,(\lambda y'(1) = 0)$

V: fixed–simply by $y(0) = y'(0) = y(1) = y''(1) = 0$

VI: free–free by $y''(0) = y'''(0) - \lambda y'(0) = y''(1) = y'''(1) - \lambda y'(1) = 0$ (3.1)

© Springer International Publishing AG, part of Springer Nature 2018
S. L. Wiggers and P. Pedersen, *Structural Stability and Vibration*, Springer Tracts
in Mechanical Engineering, https://doi.org/10.1007/978-3-319-72721-9_3

The name "simply" implies no external moment implying $y'' = 0$ and the name "translational free" implies no external shear force implying $y''' - \lambda y' = 0$, that for vibration without column load $\lambda = 0$ give $y''' = 0$.

3.2   How to Solve an Eigenvalue Problem

The homogeneous differential equation (2.12) is repeated with its non-dimensional parameters, λ for column force and ϕ for squared frequency

$$y'''' - \lambda y'' - \phi y \equiv 0$$

$$\lambda := \frac{NL^2}{EI} \quad \text{and} \quad \phi := \frac{\omega^2 L^4 \rho A}{EI} \tag{3.2}$$

This DE with chosen BC is an eigenvalue problem, that must be solved to find a non-trivial solution eigenmode (not $y(\xi) \equiv 0$), together with its corresponding eigenvalue (pair of eigenvalue) λ, ϕ. The name eigen refers to the fact that the functions y'''', y'', y should only differ by a constant (the eigenvalue). The eigen solutions are not unique.

In principle, two different approaches for solving the eigenvalue problem exist. **Find** $y(\xi) \Rightarrow \lambda, \phi$ **or find** $\lambda, \phi \Rightarrow y(\xi)$

- Find an eigenmode and use this to evaluate the corresponding eigenvalue.
- Find an eigenvalue and use this to evaluate the corresponding eigenmode.

For analytical solutions as well as numerical solutions both approaches are used and furthermore, numerical iterations are mostly necessary. With reference to large-scale numerical models this is described in Chap. 16. The present chapter focuses on analytical approaches, and with the use of analytical computer algebra the limits for analytical treatment are expanding.

3.3   Instability Modes for the Euler Cases

For some cases of DE (3.1) with BC it is not difficult to guess an eigenmode satisfying the BC, and the critical stability load λ_C is then evaluated by inserting this eigenmode in the DE $y'''' - \lambda y'' = 0$, i.e., in the present section $\phi = 0$.

3.3.1   Instability Mode for the Euler Case I

Eigenmode, i.e., instability mode or buckling mode, for Euler case I with simply–simply BC is directly seen to be $y(\xi) = C \sin(\pi \xi)$ that fulfill the BC $y(0) = y''(0) =$

$y(1) = y''(1) = 0$. Inserting $y''(\xi) = -\pi^2 y(\xi)$ and $y''''(\xi) = \pi^4 y(\xi)$ in the DE $y'''' - \lambda y'' \equiv 0$ gives the corresponding eigenvalue and thereby the critical column force, being a compression force.

Result for case I

$$\pi^4 \sin(\pi\xi) + \lambda_C \pi^2 \sin(\pi\xi) = 0 \;\Rightarrow\; \lambda_C = -\pi^2 \;\Rightarrow\; N_C = -\pi^2 \frac{EI}{L^2}$$

by definition (3.2) and the eigenmode is $y(\xi) = C \sin(\pi\xi)$ $\qquad$ (3.3)

where N_C is the critical column force.

3.3.2 Instability Modes for the Euler Cases II–IV

For Euler cases II, III, IV the two kinematic BC $y(0) = y'(0) = 0$ point towards a function of the type $y(\xi) = 1 - \cos(\alpha\pi\xi)$ with α as an unknown constant. The derivatives of this function are

$$
\begin{aligned}
y(\xi) &= C(1 - \cos(\alpha\pi\xi)) \\
y'(\xi) &= C\alpha\pi \sin(\alpha\pi\xi) \\
y''(\xi) &= C(\alpha\pi)^2 \cos(\alpha\pi\xi) \\
y'''(\xi) &= -C(\alpha\pi)^3 \sin(\alpha\pi\xi) \\
y''''(\xi) &= -C(\alpha\pi)^4 \cos(\alpha\pi\xi)
\end{aligned}
\qquad (3.4)
$$

and inserted in the DE $y'''' - \lambda y'' = 0$ the eigenvalue with the non-determined constant α is determined

$$
-C(\alpha\pi)^4 \cos(\alpha\pi\xi) - \lambda C(\alpha\pi)^2 \cos(\alpha\pi\xi) = 0 \;\Rightarrow
$$
$$
\lambda_C = -(\alpha\pi)^2 \;\Rightarrow\; N_C = -(\alpha\pi)^2 \frac{EI}{L^2}
\qquad (3.5)
$$

The not involved BC at $\xi = 1$ for the three individual Euler cases determine the specific values of the constant α.

Result for case II

$$
y(1) = y'(1) = 0 \;\Rightarrow\; \alpha = 2 \;\Rightarrow\; N_C = -4\pi^2 \frac{EI}{L^2}
$$
$$
\text{and eigenmode } y(\xi) = C(1 - \cos(2\pi\xi))
\qquad (3.6)
$$

Result for case III

$$y''(1) = 0 \;\Rightarrow\; \cos(\alpha\pi) = 0 \;\Rightarrow\; \alpha = 1/2$$

also satisfying $y'''(1) - \lambda y'(1) = 0$ by $\sin\dfrac{\pi}{2} = 0 \;\Rightarrow\; N_C = \dfrac{-\pi^2}{4}\dfrac{EI}{L^2}$

$$\text{and eigenmode } y(\xi) = C(1 - \cos(\pi\xi/2)) \quad (3.7)$$

Result for case IV

$$y'(1) = y'''(1) = 0 \;\Rightarrow\; \alpha = 1 \;\Rightarrow\; N_C = -\pi^2\dfrac{EI}{L^2}$$

$$\text{and eigenmode } y(\xi) = C(1 - \cos(\pi\xi)) \quad (3.8)$$

3.3.3   Instability Mode for the Euler Case V

An eigenmode for the critical column force is more complicated with the BC $y(1) = y''(1) = 0$ for Euler case V. Therefore, the analysis is initiated with the general solution to the DE $y'''' - \lambda y'' \equiv 0$, which includes a linear polynomial

$$
\begin{aligned}
y(\xi) &= C_1 + C_2\xi + C_3\cos(\alpha\xi) + C_4\sin(\alpha\xi) \\
y'(\xi) &= C_2 - C_3\alpha\sin(\alpha\xi) + C_4\alpha\cos(\alpha\xi) \\
y''(\xi) &= -C_3\alpha^2\cos(\alpha\xi) - C_4\alpha^2\sin(\alpha\xi) \\
y'''(\xi) &= C_3\alpha^3\sin(\alpha\xi) - C_4\alpha^3\cos(\alpha\xi) \\
y''''(\xi) &= C_3\alpha^4\cos(\alpha\xi) + C_4\alpha^4\sin(\alpha\xi)
\end{aligned}
\quad (3.9)
$$

The three BC $y(0) = y'(0) = y''(1) = 0$ can express C_1, C_2, C_3 proportional to C_4, i.e., $C_1 + C_3 = 0$, $C_2 + C_4\alpha = 0$, and $C_3\cos\alpha + C_4\sin\alpha = 0$. The fourth BC $y(1) = 0$ give $C_4(-\alpha\cos\alpha + \sin\alpha) = 0$. The constant C_4 must be nonzero to obtain a non-trivial solution. Therefore a characteristic equation, that may be solved by the Newton–Raphson method (see Chap. 16), is

$$-\alpha\cos\alpha + \sin\alpha = 0 \quad\rightarrow\quad \tan\alpha = \alpha \quad\rightarrow$$

$$\alpha_C = 4.4934\ldots \;\; (\alpha_C^2 \simeq 20.2) \quad\rightarrow\quad N_C \simeq -20.2\dfrac{EI}{L^2} \quad (3.10)$$

The nonzero value C_4 is termed C and with $C_3 = -C\tan\alpha_C$, $C_2 = -C\alpha_C\cos\alpha_C$, and $C_1 = C\sin\alpha_C$ the critical instability mode is

Result for case V

$$y(\xi) = C\,(\tan\alpha_C - \alpha_C\xi - \tan\alpha_C\cos(\alpha_C\xi) + \sin(\alpha_C\xi)) \quad (3.11)$$

Alternative general approach

As an alternative to the above analysis, a formulation in matrix algebra for the four BC may be written as a homogeneous set of linear equations

$$
\begin{bmatrix} & \text{Homogeneous part of} & \\ & \text{boundary condition (BC)} & \end{bmatrix}
\begin{Bmatrix} C_1 \\ C_2 \\ C_3 \\ C_4 \end{Bmatrix} =
\begin{Bmatrix} 0 \\ 0 \\ 0 \\ 0 \end{Bmatrix}
\tag{3.12}
$$

and a non-trivial solution $\{C_1\ C_2\ C_3\ C_4\} \neq \{0\ 0\ 0\ 0\}$ is only obtained when the determinant is zero

$$
\|[BC]\| = 0 \quad \rightarrow \quad \text{Characteristic equation} \quad \rightarrow \quad \alpha_C
\tag{3.13}
$$

Note that for critical column force only the first eigenvalues for the specific problems are presented, although higher eigenvalues mathematical exist, however not important because stability is lost at the lowest eigenvalue. For analysis of Euler cases with respect to eigenfrequencies, also the higher order eigenfrequencies are of physical importance.

Euler case VI, i.e., the free–free beam is not analyzed in relation to stability, only in relation to flexible free vibration.

3.4 Eigenfrequency Modes for the Euler Cases

The eigenmodes for the equilibrium $y'''' - \phi y \equiv 0$ without column force ($\lambda = 0$) are in most cases more complicated than the eigenmodes for the critical instability, and the general approach such as (3.13) is more relevant, except for Euler case I.

3.4.1 Eigenfrequency Modes for the Euler Case I

Results for Euler BC I

For the Euler case I of simply–simply supports it is not difficult to guess eigenmodes satisfying the BC, and the eigenvalue ϕ is then evaluated by inserting this eigenmode in the DE $y'''' - \phi y \equiv 0$. The first eigenmode for this case is identical to the instability case in Sect. 3.3.1, but with eigenmodes written for several eigenfrequencies

$$
y_n(\xi) = C \sin(n\pi\xi) \quad \text{for} \quad n = 1, 2, \dots
\tag{3.14}
$$

and the resulting eigenvalues are

$$y'''' - \phi y = 0 \quad \rightarrow \quad (n\pi)^4 C \sin(n\pi\xi) - \phi C \sin(n\pi\xi) = 0 \quad \rightarrow$$

$$\phi_n = (n\pi)^4 \quad \rightarrow \quad \omega_n^2 = (n\pi)^4 \frac{EI}{L^4 \rho A} \tag{3.15}$$

3.4.2   Eigenfrequency Modes for the Euler Cases II–VI

A convenient general expansion as used in Volterra and Zachmanoglou (1965) for these cases is

$$y(\xi) = C_1 \left(\cos(\beta\xi) + \cosh(\beta\xi)\right) + C_2 \left(\cos(\beta\xi) - \cosh(\beta\xi)\right) +$$
$$C_3 \left(\sin(\beta\xi) + \sinh(\beta\xi)\right) + C_4 \left(\sin(\beta\xi) - \sinh(\beta\xi)\right) \tag{3.16}$$

where without column force ($\lambda = 0$), the parameter β is expressed directly in the frequency parameter $\beta^4 = \phi$. The first-order derivative is

$$y'(\xi) = C_1\beta \left(-\sin(\beta\xi) + \sinh(\beta\xi)\right) + C_2\beta \left(-\sin(\beta\xi) - \sinh(\beta\xi)\right) +$$
$$C_3\beta \left(\cos(\beta\xi) + \cosh(\beta\xi)\right) + C_4\beta \left(\cos(\beta\xi) - \cosh(\beta\xi)\right)$$
$$\tag{3.17}$$

From $y(0) = 0$ follows $C_1 = 0$ and from $y'(0) = 0$ follows $C_3 = 0$ that are valid for eigenmodes of frequency for all Euler cases II–V. The simplified mode expansion with derivatives is then

$$y(\xi) = C_2 \left(\cos(\beta\xi) - \cosh(\beta\xi)\right) + C_4 \left(\sin(\beta\xi) - \sinh(\beta\xi)\right)$$
$$y'(\xi) = C_2\beta \left(-\sin(\beta\xi) - \sinh(\beta\xi)\right) + C_4\beta \left(\cos(\beta\xi) - \cosh(\beta\xi)\right)$$
$$y''(\xi) = C_2\beta^2 \left(-\cos(\beta\xi) - \cosh(\beta\xi)\right) + C_4\beta^2 \left(-\sin(\beta\xi) - \sinh(\beta\xi)\right)$$
$$y'''(\xi) = C_2\beta^3 \left(\sin(\beta\xi) - \sinh(\beta\xi)\right) + C_4\beta^3 \left(-\cos(\beta\xi) - \cosh(\beta\xi)\right) \tag{3.18}$$

The characteristic equation to determine the eigenvalues is different for the four cases II, III, IV, and V.

For Euler case II, the two BC at $\xi = 1$ are $y(1) = y'(1) = 0$ which from (3.18) imply

$$C_2 \left(\cos\beta - \cosh\beta\right) + C_4 \left(\sin\beta - \sinh\beta\right) = 0 \quad \text{and}$$
$$C_2\beta \left(-\sin\beta - \sinh\beta\right) + C_4\beta \left(\cos\beta - \cosh\beta\right) = 0 \tag{3.19}$$

The frequency equation
Zero determinant for these two linear equations is

$$(\cos \beta - \cosh \beta)(\cos \beta - \cosh \beta) - (\sin \beta - \sinh \beta)(-\sin \beta - \sinh \beta) = 0$$
$$\text{that imply} \quad \cos \beta_n \cosh \beta_n = 1 \quad \text{for} \quad n = 1, 2, \dots \tag{3.20}$$

Results for Euler BC II
Numerical solutions to equations like $\cos \beta_n \cosh \beta_n = 1$ may be determined by the
Newton–Raphson method described in Chap. 16. The corresponding eigenmodes are

$$y_n(\xi) = C \left((cos(\beta_n\xi) - cosh(\beta_n\xi)) - \frac{cos\beta_n - cosh\beta_n}{sin\beta_n - sinh\beta_n}(sin(\beta_n\xi) - sinh(\beta_n\xi)) \right)$$
$$\tag{3.21}$$

Results for Euler BC III
For Euler case III, the two BC at $\xi = 1$ are $y''(1) = y'''(1) = 0$ which from (3.18)
imply

$$C_2\beta^2 (-\cos \beta - \cosh \beta) + C_4\beta^2 (-\sin \beta - \sinh \beta) = 0 \quad \text{and}$$
$$C_2\beta^3 (\sin \beta - \sinh \beta) + C_4\beta^3 (-\cos \beta - \cosh \beta) = 0 \tag{3.22}$$

Zero determinant for these two linear equations is

$$(-\cos \beta - \cosh \beta)(-\cos \beta - \cosh \beta) - (-\sin \beta - \sinh \beta)(\sin \beta - \sinh \beta) = 0$$
$$\text{that imply} \quad \cos \beta_n \cosh \beta_n = -1 \quad \text{for} \quad n = 1, 2, \dots \tag{3.23}$$

and the corresponding eigenmodes are

$$y_n(\xi) = C \left((cos(\beta_n\xi) - cosh(\beta_n\xi)) - \frac{cos\beta_n + cosh\beta_n}{sin\beta_n + sinh\beta_n}(sin(\beta_n\xi) - sinh(\beta_n\xi)) \right)$$
$$\tag{3.24}$$

Results for Euler BC IV
For Euler case IV, the two BC at $\xi = 1$ are $y'(1) = y'''(1) = 0$ which from (3.18)
imply

$$C_2\beta (-\sin \beta - \sinh \beta) + C_4\beta (\cos \beta - \cosh \beta) = 0 \quad \text{and}$$
$$C_2\beta^3 (\sin \beta - \sinh \beta) + C_4\beta^3 (-\cos \beta - \cosh \beta) = 0 \tag{3.25}$$

Zero determinant for these two linear equations is

$$(-\sin \beta - \sinh \beta)(-\cos \beta - \cosh \beta) - (\cos \beta - \cosh \beta)(\sin \beta - \sinh \beta) = 0$$
$$\text{that imply} \quad \tan \beta_n = -\tanh \beta_n \quad \text{for} \quad n = 1, 2, \dots \tag{3.26}$$

and the corresponding eigenmodes are

$$y_n(\xi) = C\left((cos(\beta_n\xi) - cosh(\beta_n\xi)) - \frac{cos\beta_n + cosh\beta_n}{sin\beta_n + sinh\beta_n}(sin(\beta_n\xi) - sinh(\beta_n\xi))\right)$$

$$(3.27)$$

i.e., except for the numerical values of β_n are identical to (3.24) for Euler case III.

Results for Euler BC V

For Euler case V, the two BC at $\xi = 1$ are $y(1) = y''(1) = 0$ which from (3.18) imply

$$C_2(\cos\beta - \cosh\beta) + C_4(\sin\beta - \sinh\beta) = 0 \text{ and}$$

$$C_2\beta^2(-\cos\beta - \cosh\beta) + C_4\beta^2(-\sin\beta - \sinh\beta) = 0 \qquad (3.28)$$

Zero determinant for these two linear equations is

$$(\cos\beta - \cosh\beta)(-\sin\beta - \sinh\beta) - (\sin\beta - \sinh\beta)(-\cos\beta - \cosh\beta) = 0$$

$$\text{that imply} \quad \tan\beta_n = \tanh\beta_n \quad \text{for} \quad n = 1, 2, \ldots \qquad (3.29)$$

and the corresponding eigenmodes are

$$y_n(\xi) = C\left((cos(\beta_n\xi) - cosh(\beta_n\xi)) - \frac{cos\beta_n - cosh\beta_n}{sin\beta_n - sinh\beta_n}(sin(\beta_n\xi) - sinh(\beta_n\xi))\right)$$

$$(3.30)$$

i.e., except for the numerical values of β_n are identical to (3.21) for Euler case II.

Results for Euler BC VI

For Euler case VI (a free–free beam), the four BC are $y''(0) = y'''(0) = y''(1) = y'''(1) = 0$. From $y''(0) = 0$ follows $C_2 = 0$ and from $y'''(0) = 0$ follows $C_4 = 0$. The BC at $\xi = 1$ gives the same characteristic equation as for case II (a fixed–fixed beam). However, the corresponding eigenmodes are different being

$$y_n(\xi) = C\left((cos(\beta_n\xi) + cosh(\beta_n\xi)) - \frac{cos\beta_n - cosh\beta_n}{sin\beta_n - sinh\beta_n}(sin(\beta_n\xi) + sinh(\beta_n\xi))\right)$$

$$(3.31)$$

Note that the rigid eigenmodes are not included ($\omega = 0$), but they may be determined numerically.

3.5 Summary

For simple but important structural models, the critical column buckling load and the first five eigenfrequencies for transverse vibration has been determined by an analytical approach. This introductory chapter will be followed by analyzing more complicated problems such as

Extended problems

- Influence from elastic supports on column buckling loads as well as on eigenfrequencies.
- Eigenfrequencies as a function of column load.
- More general nonlinear displacement analysis for beam-columns.

To give an overview of the results in the present chapter, all the determined eigenvalues and eigenmodes are listed in Tables 3.1, 3.2, and 3.3. Numerical solutions to the characteristic equations such as shown in Table 3.2 can be obtained by the Newton–Raphson method as described in Chap. 16. The convergence of this approach is depending on the initial guess and especially with several solutions, a basic understanding of the components in a transcendental equation is valuable, say by a graphical plot of $\cos \beta$ and $\cosh \beta$ or $\tan \beta$ and $\tanh \beta$.

The content of the tables is in different forms available in a number of textbooks, such as Volterra and Zachmanoglou (1965), Rao (2007) and Shabana (1997). The presentation here of eigenmodes is chosen with largest possible uniformity.

Critical stability with $\phi = 0$

Table 3.1 Solutions for the eigenvalues that gives the critical instability load for the boundary conditions (BC) of the five Euler cases. ($\lambda_C = -\alpha^2$)

Id.	BC	Eigenmode	Eigenvalue
I	$y(0) = y''(0) =$ $y(1) = y''(1) = 0$	$y(\xi) = C \sin(\pi \xi)$	$\alpha = \pi$
II	$y(0) = y'(0) =$ $y(1) = y'(1) = 0$	$y(\xi) =$ $C(1 - \cos(2\pi \xi))$	$\alpha = 2\pi$
III	$y(0) = y'(0) =$ $y''(1) = 0$ $y'''(1) - \lambda y'(1) = 0$	$y(\xi) =$ $C(1 - \cos(\pi \xi / 2))$	$\alpha = \pi/2$
IV	$y(0) = y'(0) =$ $y'(1) = y'''(1) = 0$	$y(\xi) =$ $C(1 - \cos(\pi \xi))$	$\alpha = \pi$
V	$y(0) = y'(0) =$ $y(1) = y''(1) = 0$	$y(\xi) =$ $C(\tan \alpha - \alpha \xi -$ $\tan \alpha \cos(\alpha \xi) +$ $\sin(\alpha \xi))$	$\alpha \simeq 4.4934$

Five eigenfrequencies for $\lambda = 0$

Table 3.2 Numerical solutions for the first five eigenvalues that gives the eigenfrequencies for the boundary conditions (BC) of the six Euler cases. ($\phi = \beta^4$)

Id.	BC	Characteristic	Numerical solutions for β_{1-5}				
I	$y(0) = y''(0) = 0$ $y(1) = y''(1) = 0$	$\sin\beta = 0$	π	2π	3π	4π	5π
II	$y(0) = y'(0) = 0$ $y(1) = y'(1) = 0$	$\cos\beta\cosh\beta = 1$	4.7300	7.8532	10.9956	14.1372	17.2788
III	$y(0) = y'(0) = 0$ $y''(1) = y'''(1) = 0$	$\cos\beta\cosh\beta = -1$	1.8751	4.6941	7.8548	10.9955	14.1372
IV	$y(0) = y'(0) = 0$ $y'(1) = y'''(1) = 0$	$\tan\beta = -\tanh\beta$	2.3650	5.4978	8.6394	11.7810	14.9226
V	$y(0) = y'(0) = 0$ $y(1) = y''(1) = 0$	$\tan\beta = \tanh\beta$	3.9266	7.0686	10.2102	13.3518	16.4934
VI	$y''(0) = y'''(0) = 0$ $y''(1) = y'''(1) = 0$	$\cos\beta\cosh\beta = 1$	4.7300	7.8532	10.9956	14.1372	17.2788

Vibration eigenmodes for $\lambda = 0$

Table 3.3 Eigenmodes (vibration modes) corresponding to the Eigenvalues in Table 3.2 for the boundary conditions (BC) of the six Euler cases

Id.	Reference	Eigenmodes with β_n values in Table 3.2
I	(3.14)	$y_n(\xi) = C\sin(n\pi\xi)$
II	(3.21)	$y_n(\xi) = C\left((\cos(\beta_n\xi) - \cosh(\beta_n\xi)) - \frac{\cos\beta_n - \cosh\beta_n}{\sin\beta_n - \sinh\beta_n}(\sin(\beta_n\xi) - \sinh(\beta_n\xi))\right)$
III	(3.24)	$y_n(\xi) = C\left((\cos(\beta_n\xi) - \cosh(\beta_n\xi)) - \frac{\cos\beta_n + \cosh\beta_n}{\sin\beta_n + \sinh\beta_n}(\sin(\beta_n\xi) - \sinh(\beta_n\xi))\right)$
IV	(3.27)	$y_n(\xi) = C\left((\cos(\beta_n\xi) - \cosh(\beta_n\xi)) - \frac{\cos\beta_n + \cosh\beta_n}{\sin\beta_n + \sinh\beta_n}(\sin(\beta_n\xi) - \sinh(\beta_n\xi))\right)$
V	(3.30)	$y_n(\xi) = C\left((\cos(\beta_n\xi) - \cosh(\beta_n\xi)) - \frac{\cos\beta_n - \cosh\beta_n}{\sin\beta_n - \sinh\beta_n}(\sin(\beta_n\xi) - \sinh(\beta_n\xi))\right)$
VI	(3.21)	$y_n(\xi) = C\left((\cos(\beta_n\xi) + \cosh(\beta_n\xi)) - \frac{\cos\beta_n - \cosh\beta_n}{\sin\beta_n - \sinh\beta_n}(\sin(\beta_n\xi) + \sinh(\beta_n\xi))\right)$

Chapter 4
Beam-Columns and Applied Berry Functions

Beam-column theory is beam theory that includes bending moments from column forces. This problem is still linear in external bending loads but nonlinear in column loads for static equilibrium in displaced state. For solution of these problems, a number of functions with the common name Berry functions are presented, each of them being different combinations of trigonometric (or hyperbolic) functions. In this chapter, the external bending loads are limited to applied moments. External moments at the beam ends enable a formulation that includes rotational support.

4.1 Model of Beam-Column

Figure 4.1 shows a specific case of Fig. 2.1, with simple–simple supports and given external moments M_0, M_1. The mathematical statement of this non-homogeneous problem is

$$y'''' - \lambda y'' \equiv 0 \quad \text{with the following BC}$$

$$y(0) = y(1) = 0 \quad \text{(kinematic BC)}$$

$$y''(0) = \frac{-M_0 L^2}{EI}, \quad y''(1) = \frac{M_1 L^2}{EI} \quad \text{(static BC)} \tag{4.1}$$

according to (2.8) and (2.9) with $q = m = 0$.

The resulting end rotations are

$$\theta_0 = \left(\frac{dy}{dx}\right)_{x=0} = \frac{y'(0)}{L} \quad \text{and} \quad \theta_1 = \left(\frac{dy}{dx}\right)_{x=1} = \frac{y'(1)}{L} \tag{4.2}$$

and the goal is to determine θ_0, θ_1 as a function of M_0, M_1, with λ as parameter.

© Springer International Publishing AG, part of Springer Nature 2018

S. L. Wiggers and P. Pedersen, *Structural Stability and Vibration*, Springer Tracts in Mechanical Engineering, https://doi.org/10.1007/978-3-319-72721-9_4

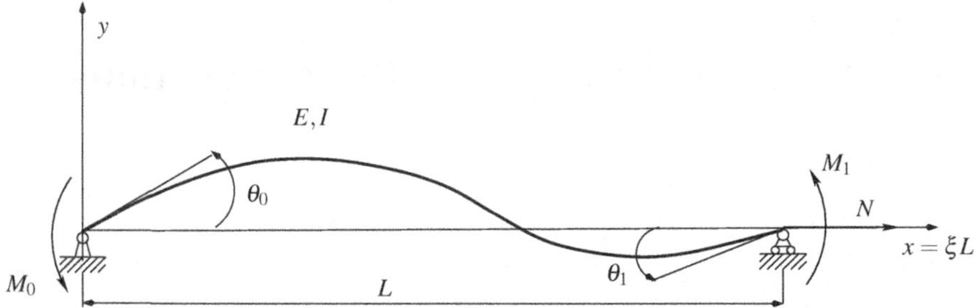

Fig. 4.1 Model of beam-column for introduction of Berry functions

Result for linear beam case

From simple beam theory without column force, the known linear solution written in matrix notation is

$$\left\{ \begin{array}{c} \theta_0 \\ \theta_1 \end{array} \right\} = \frac{L}{EI} \left[\begin{array}{cc} \frac{1}{3} & -\frac{1}{6} \\ -\frac{1}{6} & \frac{1}{3} \end{array} \right] \left\{ \begin{array}{c} M_0 \\ M_1 \end{array} \right\} \tag{4.3}$$

as found in most textbooks on beam theory, but mostly without matrix notation.

Result for nonlinear beam-column

For the DE in (4.1), part of the general solution is repeated from (3.4)

$$\begin{aligned}
y(\xi) &= C_1 + C_2\xi + C_3 \cos(\alpha\xi) + C_4 \sin(\alpha\xi) \\
y'(\xi) &= C_2 - C_3\alpha \sin(\alpha\xi) + C_4\alpha \cos(\alpha\xi) \\
y''(\xi) &= -C_3\alpha^2 \cos(\alpha\xi) - C_4\alpha^2 \sin(\alpha\xi)
\end{aligned} \tag{4.4}$$

where $\alpha = \sqrt{-\lambda}$. The actual BC in matrix notation are

$$\left[\begin{array}{cccc} 1 & 0 & 1 & 0 \\ 1 & 1 & \cos\alpha & \sin\alpha \\ 0 & 0 & -\alpha^2 & 0 \\ 0 & 0 & -\alpha^2\cos\alpha & -\alpha^2\sin\alpha \end{array} \right] \left\{ \begin{array}{c} C_1 \\ C_2 \\ C_3 \\ C_4 \end{array} \right\} = \frac{L^2}{EI} \left\{ \begin{array}{c} 0 \\ 0 \\ -M_0 \\ M_1 \end{array} \right\} \tag{4.5}$$

that are solved with respect to the constants C_1, C_2, C_3, C_4 to obtain

$$y(\xi) = \frac{L^2}{\alpha^2 EI} \left((\cos(\alpha\xi) - \cot\alpha\sin(\alpha\xi) + \xi - 1)M_0 + (\xi - \frac{\sin(\alpha\xi)}{\sin\alpha})M_1 \right)$$

$$y'(\xi) = \frac{L^2}{\alpha^2 EI} \left((-\alpha\sin(\alpha\xi) - \alpha\cot\alpha\cos(\alpha\xi) + 1)M_0 + (1 - \frac{\alpha\cos(\alpha\xi)}{\sin\alpha})M_1 \right)$$

$$\tag{4.6}$$

The end rotations are

$$
\theta_0 = \frac{y'(0)}{L} = \frac{C_2 + C_4\alpha}{L} \quad \text{and}
$$
$$
\theta_1 = \frac{y'(1)}{L} = \frac{C_2 - C_3\alpha \sin\alpha + C_4\alpha \cos\alpha}{L} \tag{4.7}
$$

that in matrix notation shows the final result

$$
\begin{Bmatrix} \theta_0 \\ \theta_1 \end{Bmatrix} = \frac{L}{EI} \begin{bmatrix} B_1 & B_0 \\ B_0 & B_1 \end{bmatrix} \begin{Bmatrix} M_0 \\ M_1 \end{Bmatrix} \tag{4.8}
$$

where the Berry functions B_0, B_1 are defined by

$$
B_0 := \frac{1}{\alpha^2}\left(1 - \frac{\alpha}{\sin\alpha}\right) \quad \text{and} \quad B_1 := \frac{1}{\alpha^2}(1 - \alpha \cot\alpha) \tag{4.9}
$$

with the limiting values for $\lambda = 0$ being $B_0 = -\frac{1}{6}$ and $B_1 = \frac{1}{3}$ in agreement with the linear beam result (4.3). In Sect. 4.2, the determination of the limiting values is discussed, and see the footnote.[1]

Figure 4.2 shows for the non-dimensional column force interval $-50 < -\lambda < 150$, a display of B_0 and B_1 together with the determinant of the 2×2 matrix in (4.8), i.e., $B_1^2 - B_0^2$. From (4.8), it is seen that for zero value of this determinant the solution θ_0, θ_1 may be infinite for any combination of the external loads M_0, M_1, i.e., in a critical state of stability. Further displays of Berry functions are restricted to the interval $-10 < -\lambda < 40$.

---

[1] For functions f and g which are differentiable on an open interval I except possibly at a point c contained in I, with

$$
\lim_{x \to c} f(x) = 0 \quad \text{and} \quad \lim_{x \to c} g(x) = 0 \tag{4.10}
$$

the general form of L'Hospital's rule states

$$
\lim_{x \to c} \frac{f(x)}{g(x)} = \lim_{x \to c} \frac{f'(x)}{g'(x)} \tag{4.11}
$$

if $g'(x)$ for $x = c$ exist.

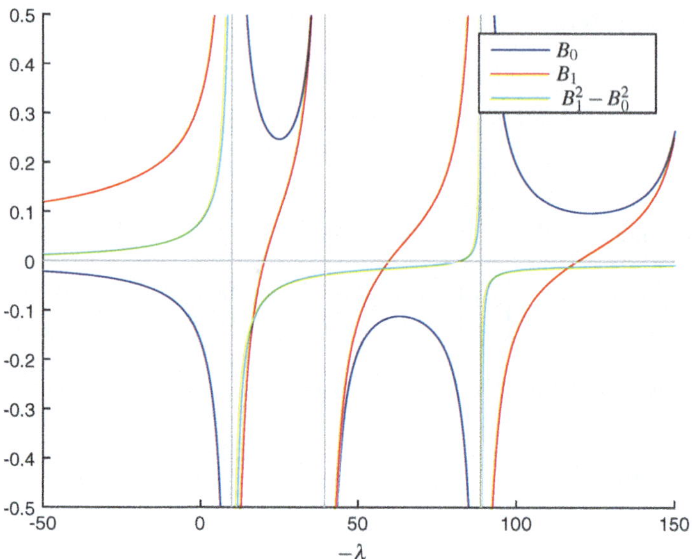

Fig. 4.2 Berry functions B_0 and B_1 as a function of the non-dimensional column force. The determinant $B_1^2 - B_0^2$ is also shown

4.2   General Moment Loads

An important conclusion from the analysis above is that the resulting displacements (end rotations) in (4.8) are linear in the beam loads M_0, M_1. This linear dependence on beam loads is general, and thus, superposition is possible, assuming that the column force is unchanged. Results are nonlinear in the column force, described non-dimensional by λ ($\alpha = \sqrt{-\lambda} = i\sqrt{\lambda}$).

Matching BC

An external moment acting at the relative position η is shown in the model in Fig. 4.3. Solution to this problem is obtained by matching the BC for the two beams, I to the left of the acting external moment and II to the right in Fig. 4.3. The necessary equations are

$$y_I'''' - \lambda_I y_I'' \equiv 0 \quad \text{with} \quad \lambda_I = \frac{NL^2}{EI}(1 - \eta)^2 \quad \text{and}$$

$$y_{II}'''' - \lambda_{II} y_{II}'' \equiv 0 \quad \text{with} \quad \lambda_{II} = \frac{NL^2}{EI}\eta^2$$

and kinematic/static BC: $y_I(0) = y_I''(0) = y_{II}(1) = y_{II}''(1) = 0$

continuity kinematic BC: $y_I(1) = y_{II}(0)$ and $\dfrac{y_I'(1)}{1 - \eta} = \dfrac{y_{II}'(0)}{\eta}$

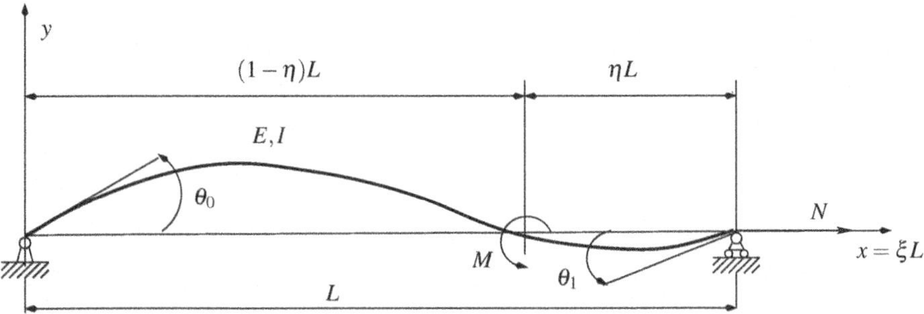

Fig. 4.3 Model of beam-column with arbitrary position η of the moment beam load

$$\text{static moment BC:} \quad \frac{y_I''(1)}{(1-\eta)^2} - \frac{y_{II}''(0)}{\eta^2} = \frac{ML^2}{EI}$$

$$\text{static shear force BC:} \quad -\frac{y_I'''(1) - \lambda_I y_I'(1)}{(1-\eta)^3} + \frac{y_{II}'''(0) - \lambda_{II} y_{II}'(0)}{\eta^3} = 0 \quad (4.12)$$

with last conditions from (2.9).

Resulting displacements

The solution obtained to the problem (4.12) is

$$y_\eta(\xi) = \frac{ML^2}{EI} \left(\frac{\xi \sin\alpha - \cos(\alpha\eta)\sin(\alpha\xi)}{\alpha^2 \sin\alpha} \right) \quad \text{for } \xi < 1 - \eta \quad (4.13)$$

with end rotations

$$\begin{Bmatrix} \theta_0 \\ \theta_1 \end{Bmatrix} = \frac{ML}{EI} \begin{Bmatrix} B_\eta \\ B_{1-\eta} \end{Bmatrix} \quad (4.14)$$

The solution for $\xi > 1 - \eta$ is found from (4.13) by the following substitutions: $\xi \hookrightarrow 1 - \xi$; $\eta \hookrightarrow 1 - \eta$ and $y_\eta(\xi) \hookrightarrow -y_\eta(\xi)$.

Basic Berry function

The defined Berry function is

$$\begin{aligned} B_\eta &:= \frac{\alpha\cos(\eta\alpha) - \sin\alpha}{-\alpha^2 \sin\alpha} \\ &= \frac{\sqrt{-\lambda}\cos(\eta\sqrt{-\lambda}) - \sin\sqrt{-\lambda}}{\lambda\sin\sqrt{-\lambda}} = \frac{\sqrt{\lambda}\cosh(\eta\sqrt{\lambda}) - \sinh\sqrt{\lambda}}{\lambda\sinh\sqrt{\lambda}} \end{aligned} \quad (4.15)$$

where the relations $\cos(iz) = \cosh(z)$ and $\sin(iz) = i\sinh(z)$ are applied such that practical formula is available for compression ($\lambda < 0$) as well as for tension ($\lambda > 0$). The Berry function B_η relates to an external moment at position η. The limiting

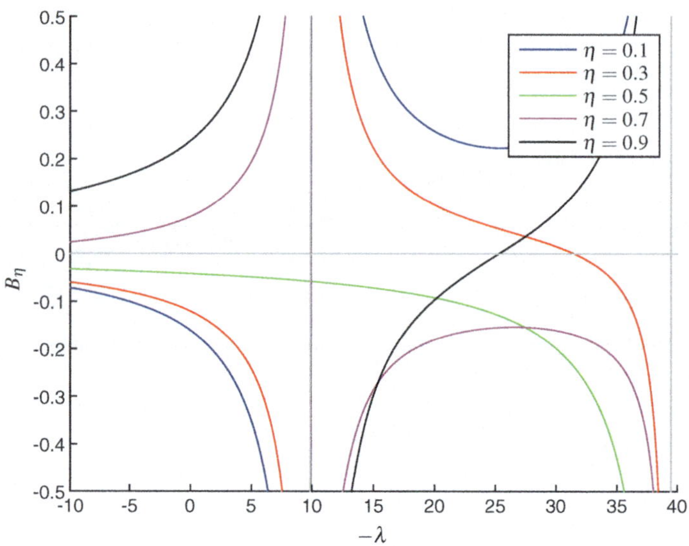

Fig. 4.4 Berry functions B_η as a function of the non-dimensional column force, with the position of the external beam moment as parameter

values $B_\eta(0)$ without column force are $B_\eta(0) = \frac{\eta^2}{2} - \frac{1}{6}$ in agreement with the values in (4.3). The determination of $B_\eta(0)$ from (4.15) is not so direct, because nominator as well as denominator up to the third order derivative is zero, so L'Hospital's rule has to be applied more than once. Figure 4.4 shows examples of $B_\eta(\lambda)$ with parameter $0 < \eta < 1$. For the specific cases of end moment loads, the Berry functions B_0 for $\eta = 0$ and B_1 for $\eta = 1$ are presented in (4.9) in an extended domain.

As superposition of results from different beam loads is possible, then distributed moment beam loads can be analyzed by integration of Berry functions.

4.3 Elastic Support Against End Rotations

The treated case with result (4.13) may be used to obtain results for other BC than simple–simple supports. Figure 4.5 shows BC with linear elastic rotational springs at the ends. The definition of spring stiffness S_0, S_1 is illustrated physically as well as with equivalent reaction moments $M_0 = -S_0\theta_0$, $M_1 = -S_1\theta_1$.

The analysis of this case concentrates on the resulting end rotations θ_0, θ_1 that may be obtained by superposition of three external moments, i.e., directly from (4.8) for $M_0 = -S_0\theta_0$ and $M_1 = -S_1\theta_1$ and from (4.13) for M

$$\begin{Bmatrix} \theta_0 \\ \theta_1 \end{Bmatrix} = \frac{L}{EI} \begin{bmatrix} B_1 & B_0 \\ B_0 & B_1 \end{bmatrix} \begin{Bmatrix} M_0 \\ M_1 \end{Bmatrix} + \frac{ML}{EI} \begin{Bmatrix} B_\eta \\ B_{1-\eta} \end{Bmatrix} \tag{4.16}$$

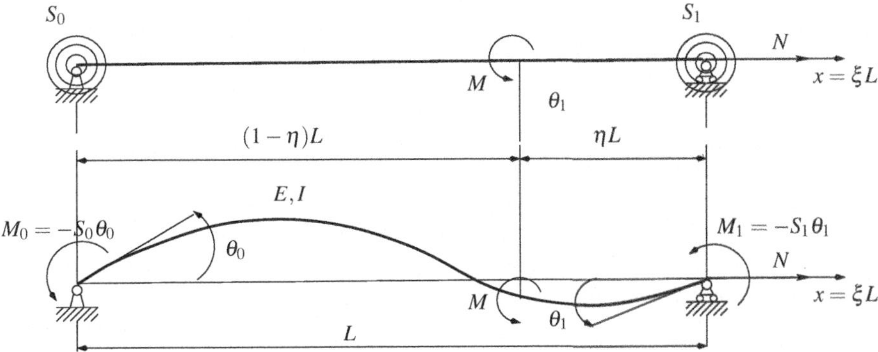

Fig. 4.5 Beam-column with linear elastic rotational end springs, and below the equivalent reaction moments M_0, M_1

Substituting $M_0 = -S_0\theta_0$ and $M_1 = -S_1\theta_1$ and the non-dimensional spring stiffnesses defined by

$$s_0 := \frac{LS_0}{EI} \quad \text{and} \quad s_1 := \frac{LS_1}{EI} \tag{4.17}$$

gives

$$\begin{bmatrix} 1 + s_0 B_1 & s_1 B_0 \\ s_0 B_0 & 1 + s_1 B_1 \end{bmatrix} \begin{Bmatrix} \theta_0 \\ \theta_1 \end{Bmatrix} = \frac{ML}{EI} \begin{Bmatrix} B_\eta \\ B_{1-\eta} \end{Bmatrix} \tag{4.18}$$

and from there the solution

$$\begin{Bmatrix} \theta_0 \\ \theta_1 \end{Bmatrix} = \frac{1}{D} \frac{ML}{EI} \begin{Bmatrix} B_\eta(1 + s_1 B_1) - B_{1-\eta} s_1 B_0 \\ -B_\eta s_0 B_0 + B_{1-\eta}(1 + s_0 B_1) \end{Bmatrix} \tag{4.19}$$

with determinant

$$D = 1 + (s_1 + s_0) B_1 + s_1 s_0 (B_1^2 - B_0^2) \tag{4.20}$$

As discussed in relation to (4.8), the case of zero determinant $D = 0$ is important for stability analysis, but for the nonlinear analysis it is assumed that $D \neq 0$ and a solution to (4.19) exist.

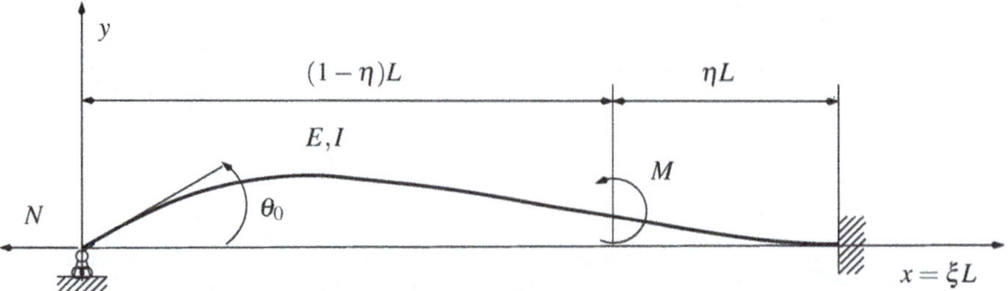

Fig. 4.6 Beam-column with simply fixed BC, i.e., a Euler case V

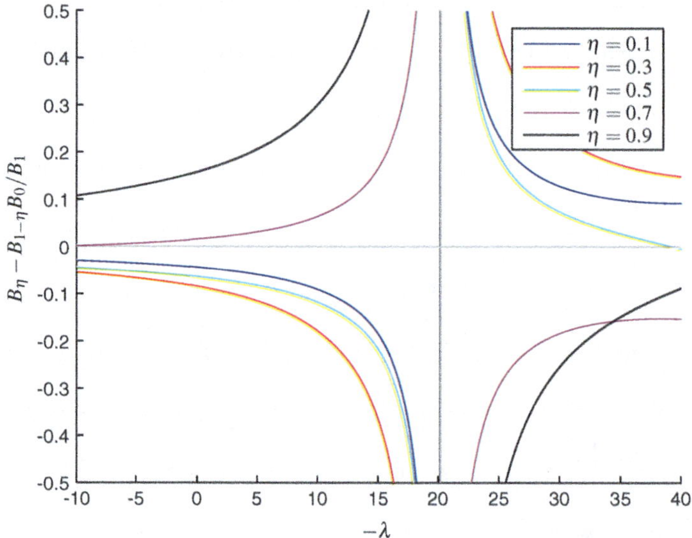

Fig. 4.7 Combined Berry functions as a function of the non-dimensional column force, with the position of the external beam moment as parameter. The actual combination relates to the specific BC in Fig. 4.6

4.3.1 A Fixed Support as a Limiting Case

The specific support with the springs $s_0 = 0, s_1 = \infty$ corresponds to the case in Fig. 4.6. Only θ_0 is of interest as $\theta_1 = 0$, and only components in (4.19) with s_1 as factor remain and give

$$\theta_0 = \frac{ML}{EI}(B_\eta - B_{1-\eta}B_0/B_1) \;\text{ with }\; \theta_0(\lambda = 0) = \frac{ML}{EI}\frac{\eta}{4}(3\eta - 2) \qquad (4.21)$$

The specific combination of Berry functions in (4.21) is graphically shown in Fig. 4.7. Note the singularity at $\lambda \simeq 20.2$, where $B_1 = 0$. This singularity holds for any position of the external moment M and corresponds to the instability force for Euler case V in Chap. 3.

Chapter 5
Shear Beam Loads and Cantilever Beam-Columns

A basic problem with external beam shear load Q at a given position has close similarities to the case of external moments in this chapter, but the involved Berry functions (now with bar notation) are more complicated. In addition to rotational spring supports at the beam-column ends, translational spring support is included at the right beam end. Results for cantilever beam-columns are then available, then with a number of parameters: load(s) position η, rotational spring stiffnesses S_0, S_1 (dimensionless s_0, s_1), and translational spring stiffness K (dimensionless k).

5.1 Shear Loads on Beam-Columns

The results in Sect. 4.2 gave the possibility to evaluate the beam displacement for any moment load. The analysis of the case in Fig. 5.1 gives the basis to analyze for any shear force load.

Figure 5.1 shows a beam-column case with simple–simple supports and given external shear force Q acting at the relative position η.

Resulting displacements

The general solution for each of the two beam-columns is applied with BC as in (4.12), except that now $M = 0$ and $Q \neq 0$. The final result may also be found in Timoshenko and Gere (1961) and is

$$y_\eta(\xi) = \frac{QL^3}{EI} \left(\frac{\sin(\eta\alpha)\sin(\alpha\xi) - \eta\alpha\xi\sin\alpha}{\alpha^3\sin\alpha} \right) \quad \text{for } \xi < 1 - \eta \qquad (5.1)$$

with $\alpha = \sqrt{-\lambda} = i\sqrt{\lambda}$ and end rotations

© Springer International Publishing AG, part of Springer Nature 2018
S. L. Wiggers and P. Pedersen, *Structural Stability and Vibration*, Springer Tracts
in Mechanical Engineering, https://doi.org/10.1007/978-3-319-72721-9_5

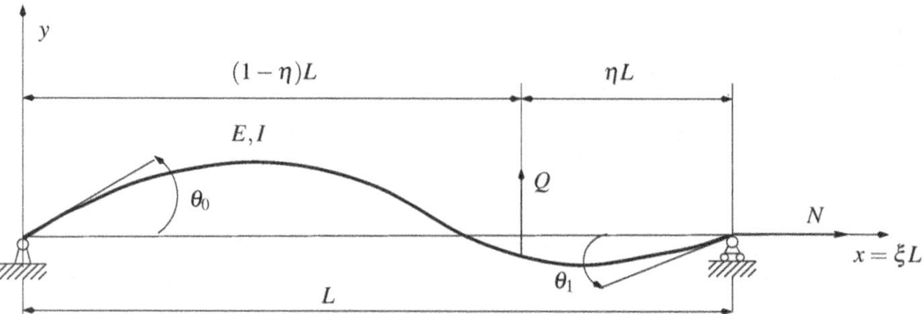

Fig. 5.1 Beam-column with general position of an external shear force Q

$$\begin{Bmatrix} \theta_0 \\ \theta_1 \end{Bmatrix} = \frac{QL^2}{EI} \begin{Bmatrix} \bar{B}_\eta \\ -\bar{B}_{1-\eta} \end{Bmatrix} \tag{5.2}$$

The solution for $\xi > 1 - \eta$ is found from (5.1) by the following substitutions: $\xi \hookrightarrow 1 - \xi$ and $\eta \hookrightarrow 1 - \eta$.

New Berry function

The defined new Berry function (with added bar notation) is

$$\begin{aligned} \bar{B}_\eta &:= \frac{\eta \sin \alpha - \sin(\eta \alpha)}{-\alpha^2 \sin \alpha} \\ &= \frac{\eta \sin \sqrt{-\lambda} - \sin(\eta\sqrt{-\lambda})}{\lambda \sin \sqrt{-\lambda}} = \frac{\eta \sinh \sqrt{\lambda} - \sinh(\eta\sqrt{\lambda})}{\lambda \sinh \sqrt{\lambda}} \end{aligned} \tag{5.3}$$

with $\bar{B}_\eta(\lambda = 0) = \frac{1}{6}\eta(1 - \eta^2)$.

Figure 5.2 shows a graphical display of this new Berry function with the relative position η of the shear force Q as parameter. Note the singular behavior at $\lambda = -\pi^2$ (and $\lambda = -4\pi^2$) that also where the case in Fig. 4.4.

Note in (5.1), the different signs for θ_0 and θ_1 which were not the case in (4.13) seen to agree by rotating the problems 180 degrees such that the load Q changes direction in contrast to the moment M.

Distributed external shear forces

A beam-column with several external shear forces for the same column force is analyzed by superposition. For continuously distributed external shear forces, integration of $\bar{B}_\eta$ may be needed, and we list the necessary formula to analyze uniform distributed and linear varying shear forces

$$\int_0^1 \bar{B}_\eta d\eta = \frac{1 - \cos \alpha - \alpha \sin \alpha / 2}{\alpha^3 \sin \alpha} \tag{5.4}$$

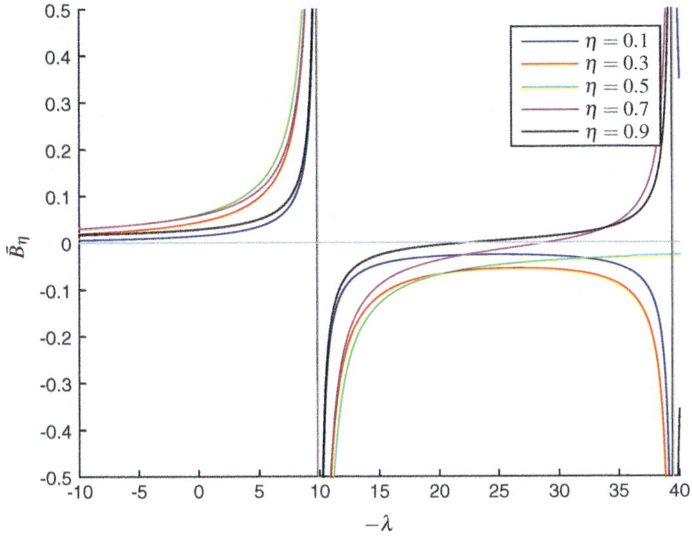

Fig. 5.2 Berry functions $\bar{B}_\eta$ as a function of the non-dimensional column force, with the position η of the external beam shear force as parameter

and

$$\int_0^1 \eta \bar{B}_\eta d\eta = \frac{-\alpha \cos\alpha + (1 - \alpha^2/3)\sin\alpha}{\alpha^4 \sin\alpha} \tag{5.5}$$

Rotational springs

Without repeating the analysis from Chap. 4, the solution for a beam-column which is elastically supported at the ends is

$$\begin{Bmatrix} \theta_0 \\ \theta_1 \end{Bmatrix} = \frac{1}{D} \frac{QL^2}{EI} \begin{Bmatrix} \bar{B}_\eta(1 + s_1 B_1) + \bar{B}_{1-\eta} s_1 B_0 \\ -\bar{B}_\eta s_0 B_0 - \bar{B}_{1-\eta}(1 + s_0 B_1) \end{Bmatrix} \tag{5.6}$$

with determinant

$$D = 1 + (s_1 + s_0)B_1 + s_1 s_0 (B_1^2 - B_0^2) \tag{5.7}$$

5.1.1   A Fixed Support as a Limiting Case

Similar to Sect. 4.3.1, the specific support with the springs $s_0 = 0, s_1 = \infty$ corresponds to the case in Fig. 4.6, now with beam load force Q. Only θ_0 is of interest as $\theta_1 = 0$, and only components in (5.6) with s_1 as factor remain and give

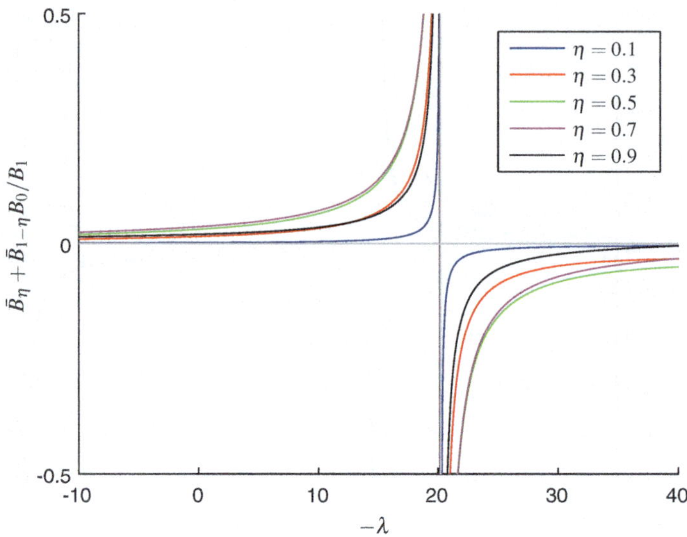

Fig. 5.3 Combined Berry functions as a function of the non-dimensional column force, with the position η of the external force Q as parameter. The actual combination relates to the specific BC in Fig. 4.6, now with beam load force Q

$$\theta_0 = \frac{QL^2}{EI}(\bar{B}_\eta + \bar{B}_{1-\eta}B_0/B_1) \;\;\text{with}\;\; \theta_0(\lambda = 0) = \frac{QL^2}{EI}\frac{\eta^2}{4}(1-\eta) \qquad (5.8)$$

The function $(\bar{B}_\eta + \bar{B}_{1-\eta}B_0/B_1)$ as a function of λ is graphically illustrated in Fig. 5.3 with the singularity at $\lambda \simeq 20.2$.

5.2   Cantilever Beam-Columns

The nonlinear analysis is extended to include cases with translational displacements at the right end, such as cantilever with free end or with translational spring support. The cantilever case includes a translational spring with stiffness K that is described non-dimensional by k

$$k = \frac{L^3 K}{EI} \qquad (5.9)$$

Figure 5.4 shows the actual physical model with three springs as well as the mathematical equivalent model with end displacement parameters $\theta_0.\theta_1,\,\psi$.

Moment equilibrium for the model in Fig. 5.4 is

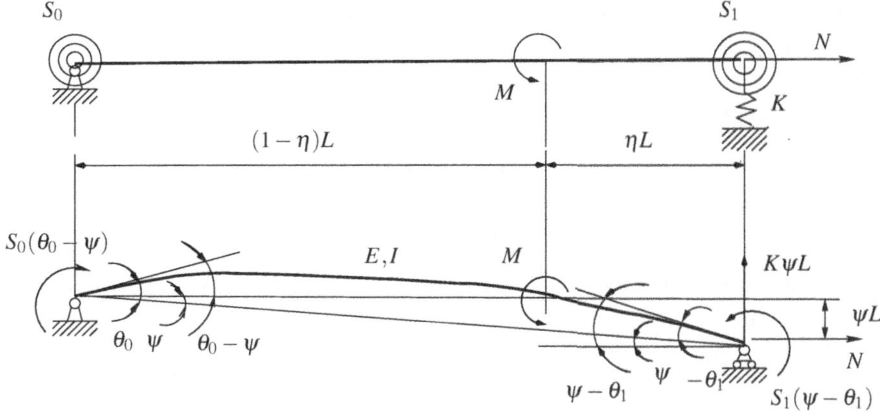

Fig. 5.4 Cantilever beam-column with translational spring at right end. Note, θ_0 and θ_1 are measured from the line between the displaced ends

$$S_0(\theta_0 - \psi) = M + N\psi L + K\psi L^2 + S_1(\psi - \theta_1) \tag{5.10}$$

or in non-dimensional formulation

$$ML/(EI) + \lambda\psi + k\psi + s_1(\psi - \theta_1) - s_0(\theta_0 - \psi) = 0 \tag{5.11}$$

Using equation (4.16) in Sect. 4.3 for this model gives

$$\begin{Bmatrix} \theta_0 \\ \theta_1 \end{Bmatrix} = \frac{L}{EI} \begin{bmatrix} B_1 & B_0 \\ B_0 & B_1 \end{bmatrix} \begin{Bmatrix} -S_0(\theta_0 - \psi) \\ S_1(\psi - \theta_1) \end{Bmatrix} + \frac{ML}{EI} \begin{Bmatrix} B_\eta \\ B_{1-\eta} \end{Bmatrix} \tag{5.12}$$

and the result (extended relative to (4.18)) in matrix notation is

$$\begin{bmatrix} 1 + s_0 B_1 & s_1 B_0 & -s_0 B_1 - s_1 B_0 \\ s_0 B_0 & 1 + s_1 B_1 & -s_0 B_0 - s_1 B_1 \\ s_0 & s_1 & -k - \lambda - s_1 - s_0 \end{bmatrix} \begin{Bmatrix} \theta_0 \\ \theta_1 \\ \psi \end{Bmatrix} = \frac{ML}{EI} \begin{Bmatrix} B_\eta \\ B_{1-\eta} \\ 1 \end{Bmatrix} \tag{5.13}$$

Solution for general spring supports

The solution to this system of equations is

$$\begin{Bmatrix} \theta_0 \\ \theta_1 \\ \psi \end{Bmatrix} = \frac{1}{D} \frac{ML}{EI} [F] \begin{Bmatrix} B_\eta \\ B_{1-\eta} \\ 1 \end{Bmatrix} \tag{5.14}$$

with definition of determinant D and the matrix $[F]$

$$D := (k + \lambda)(1 + (s_1 + s_0)B_1 + s_1 s_0(B_1^2 - B_0^2)) + 2s_1 s_0(B_1 - B_0) + (s_1 + s_0)$$

$$[F] := \begin{bmatrix} f_{11} & f_{12} & f_{13} \\ f_{21} & f_{22} & f_{23} \\ f_{31} & f_{32} & f_{33} \end{bmatrix} \quad \text{with the components being}$$

$$f_{11} = (k + \lambda)(1 + s_1 B_1) + s_1 + s_0 + s_1 s_0(B_1 - B_0)$$

$$f_{22} = (k + \lambda)(1 + s_0 B_1) + s_1 + s_0 + s_1 s_0(B_1 - B_0)$$

$$f_{33} = -1 - (s_1 + s_0)B_1 - s_1 s_0(B_1^2 - B_0^2))$$

$$f_{12} = -(k + \lambda)s_1 B_0 + s_1 s_0(B_1 - B_0), \quad f_{21} = -(k + \lambda)s_0 B_0 + s_1 s_0(B_1 - B_0)$$

$$f_{13} = -s_1 B_0 - s_0 B_1 - s_1 s_0(B_1^2 - B_0^2), \quad f_{31} = s_0 + s_1 s_0(B_1 - B_0)$$

$$f_{23} = -s_0 B_0 - s_1 B_1 - s_1 s_0(B_1^2 - B_0^2), \quad f_{32} = s_1 + s_1 s_0(B_1 - B_0) \tag{5.15}$$

Note that for $k = \infty$, then $\psi = 0$ and (5.15) is equal to (4.19). The results (5.6), (5.14) and (5.15) are applied in Chaps. 7 and 8.

5.2.1   Two Cantilever Cases

The pure cantilever case of $k = s_1 = 0$ and $s_0 = \infty$ results for a single moment load M at the position η in Fig. 5.5

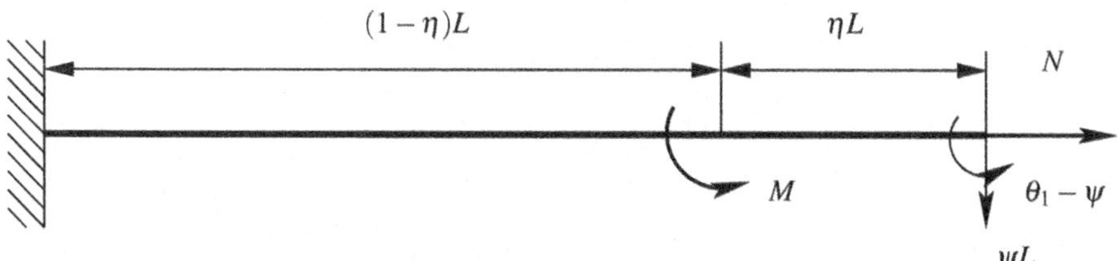

Fig. 5.5 Cantilever beam-column with a single moment beam load

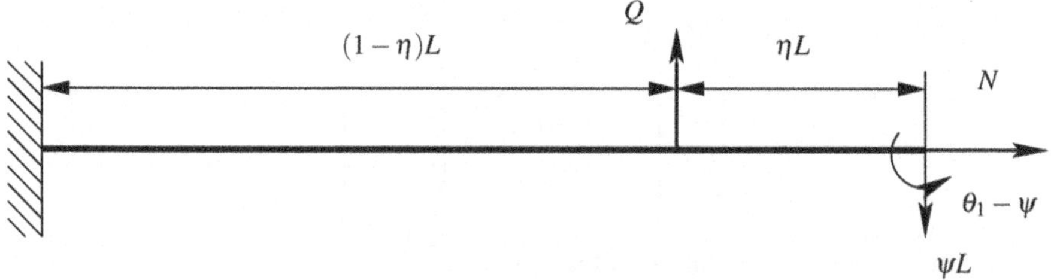

Fig. 5.6 Cantilever beam-column with a single shear force beam load

$$\left\{ \begin{matrix} \theta_0 = \psi \\ \theta_1 \end{matrix} \right\} = \frac{ML}{EI} \frac{1}{\lambda B_1 + 1} \left\{ \begin{matrix} B_\eta - B_1 \\ \lambda(B_1 B_{1-\eta} - B_0 B_\eta) + B_{1-\eta} - B_0 \end{matrix} \right\} \qquad (5.16)$$

The similar case with a transverse shear load Q at the position η results in Fig. 5.6

$$\left\{ \begin{matrix} \theta_0 = \psi \\ \theta_1 \end{matrix} \right\} = \frac{QL^2}{EI} \frac{1}{\lambda B_1 + 1} \left\{ \begin{matrix} \bar{B}_\eta - B_1(1 - \eta) \\ -\lambda(B_1 \bar{B}_{1-\eta} + B_0 \bar{B}_\eta) - \bar{B}_{1-\eta} - B_0(1 - \eta) \end{matrix} \right\}$$
$$(5.17)$$

Chapter 6
Beam-Column Eigenfrequencies

In Sect. 3.3, the DE for eigenfrequencies is derived and discussed in relation to simply supported BC. Now including rotational end springs to get further BC, the analytical solutions complicate and generalized basic Berry functions with an added tilde notation must be introduced. For each case of no axial force, eigenfrequencies as a function of conservative axial force are obtained. Graphically results are shown for different combinations of rotational spring support. It is noted that the DE may describe the combination of harmonic vibration with continuous support (Winkler support) and with linear magnetic attraction. Relations to Kolousek functions are pointed out.

6.1 Mathmatical Model for Different Physical Problems

Initially in Chap. 2, the full DE for uniform beam-column (constant EI) was derived with non-dimensional column force parameter λ and non-dimensional squared frequency parameter ϕ, being (repeating)

$$y'''' - \lambda y'' - \phi y \equiv 0 \tag{6.1}$$

and in the earlier chapters solved for pure stability eigenvalue problems ($\phi = 0$), pure eigenfrequency problems ($\lambda = 0$), and for nonlinear beam-column problems with different beam loads and different BC. In the present chapter, the full DE (6.1) is analyzed and the parameter ϕ describes physical different problems in addition to squared frequency. Limiting the presentation to simply supported BC with given external end moments M_0, M_1, that might also describe elastic rotational springs, Fig. 6.1 shows three structural models. The actual BC for these models are

© Springer International Publishing AG, part of Springer Nature 2018
S. L. Wiggers and P. Pedersen, *Structural Stability and Vibration*, Springer Tracts
in Mechanical Engineering, https://doi.org/10.1007/978-3-319-72721-9_6

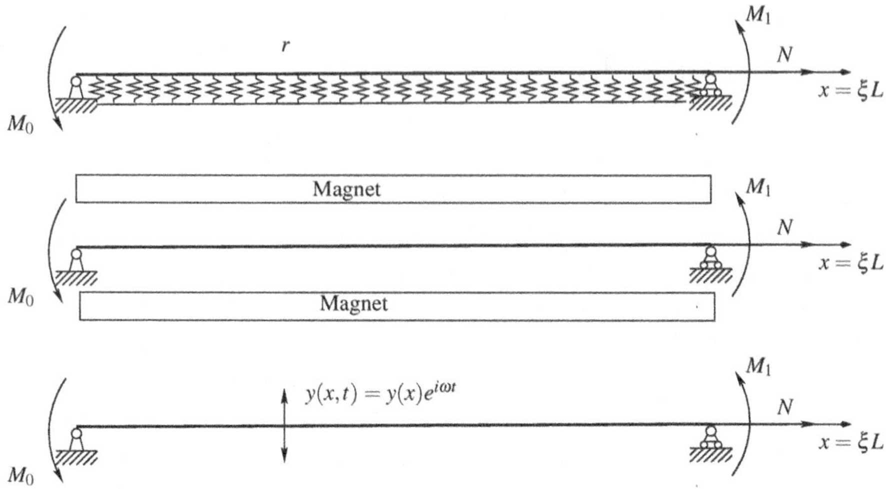

Fig. 6.1 Three different physical problems with same mathematical model by DE (6.1) and BC (6.2)

$$y(0) = y(1) = 0 \quad y''(0) = \frac{-M_0 L^2}{EI} \quad y''(1) = \frac{M_1 L^2}{EI} \tag{6.2}$$

The upper model in Fig. 6.1 has continuously linear elastic support, often termed Winkler supports, and for this model, ϕ is negative

$$\phi < 0 \quad \text{for Winkler support} \tag{6.3}$$

The model in the middle in Fig. 6.1 has continuously linear magnetic attraction and corresponds to a support with negative springs. For this case, ϕ is positive

$$\phi > 0 \quad \text{for linear magnetic attraction} \tag{6.4}$$

Finally, the lower model in Fig. 6.1 refers to a static formulation of a vibration by d'Alembert principle as shown in Sect. 2.3. For this case, ϕ is positive

$$\phi > 0 \quad \text{for modeling harmonic vibration} \tag{6.5}$$

6.2   Solution of DE (6.1) with BC (6.2)

The general solution to DE (6.1) is given with four unknown constants C_1, C_2, C_3, C_4, here different from the alternative in (3.16). The derivatives to be used are given by

$$y(\xi) = C_1 \cosh(\alpha\xi) + C_2 \sinh(\alpha\xi) + C_3 \cos(\beta\xi) + C_4 \sin(\beta\xi)$$
$$y'(\xi) = C_1\alpha \sinh(\alpha\xi) + C_2\alpha \cosh(\alpha\xi) - C_3\beta \sin(\beta\xi) + C_4\beta \cos(\beta\xi)$$
$$y''(\xi) = C_1\alpha^2 \cosh(\alpha\xi) + C_2\alpha^2 \sinh(\alpha\xi) - C_3\beta^2 \cos(\beta\xi) - C_4\beta^2 \sin(\beta\xi)$$
$$y'''(\xi) = C_1\alpha^3 \sinh(\alpha\xi) + C_2\alpha^3 \cosh(\alpha\xi) + C_3\beta^3 \sin(\beta\xi) - C_4\beta^3 \cos(\beta\xi)$$
$$y''''(\xi) = C_1\alpha^4 \cosh(\alpha\xi) + C_2\alpha^4 \sinh(\alpha\xi) + C_3\beta^4 \cos(\beta\xi) + C_4\beta^4 \sin(\beta\xi)$$
$$\tag{6.6}$$

Inserting y, y'', y'''' from (6.6) in the DE (6.1) give the relations

$$\alpha = \sqrt{\frac{\lambda}{2} + \sqrt{\frac{\lambda^2}{4} + \phi}} \quad \text{and} \quad \beta = \sqrt{-\frac{\lambda}{2} + \sqrt{\frac{\lambda^2}{4} + \phi}}$$
$$\lambda = \alpha^2 - \beta^2 \quad \text{and} \quad \phi = \alpha^2\beta^2 \tag{6.7}$$

and the specific BC of simply—simply supports results in the solution

$$y(\xi) = \frac{1}{D}\frac{L^2}{EI}((-\sinh\alpha \sin\beta \cosh(\alpha\xi) + \cosh\alpha \sin\beta \sinh(\alpha\xi) +$$
$$\sinh\alpha \sin\beta \cos(\beta\xi) - \sinh\alpha \cos\beta \sin(\beta\xi))M_0 +$$
$$(\sin\beta \sinh(\alpha\xi) - \sinh\alpha \sin(\beta\xi))M_1) \quad \text{with}$$
$$D := (\alpha^2 + \beta^2)\alpha \sin\beta \tag{6.8}$$

Note that combinations of trigonometric functions may in numerical evaluation give rise to unexpected problems. It is mentioned here because a result like (6.8) may require good number representation to get a reliable evaluation.

Basis Berry functions

As in the earlier chapters, the end rotations θ_0, θ_1 as a function of the end moments M_0, M_1, and the axial load λ is of main interest, as in (4.8). The extended formulation when including the parameter ϕ involves generalized Berry functions with notation $\widetilde{B}_0$, $\widetilde{B}_1$, with notation added a tilde, and for $\phi = 0$, these functions are the basis Berry functions

$$B_0(\lambda) = \widetilde{B}_0(\lambda, \phi = 0) \quad \text{and} \quad B_1(\lambda) = \widetilde{B}_1(\lambda, \phi = 0) \tag{6.9}$$

Written in matrix notation, using differentiation of the result in (6.8) and $\theta_0 = y'(0)/L$ and $\theta_1 = y'(1)/L$, the result is

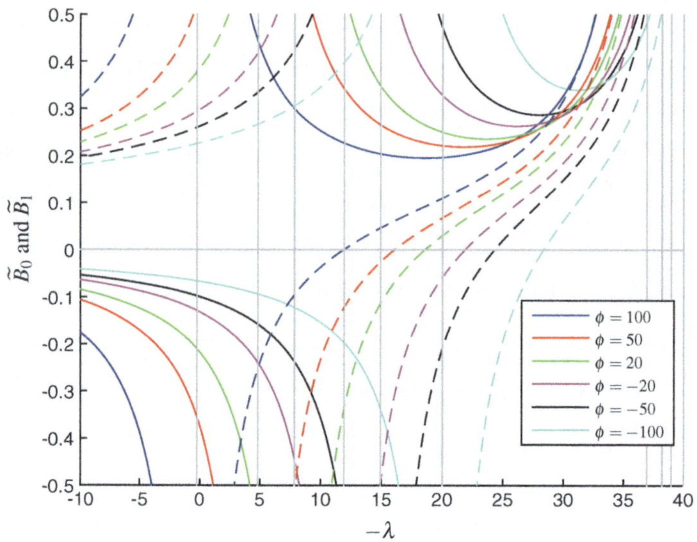

Fig. 6.2 Generalized Berry functions $\widetilde{B}_0$ (full lines, left starting negative) and $\widetilde{B}_1$ (dashed lines, left starting positive) as a function of non-dimensional column force λ with ϕ as parameter

$$\begin{Bmatrix} \theta_0 \\ \theta_1 \end{Bmatrix} = \frac{L}{EI} \begin{bmatrix} \widetilde{B}_1 & \widetilde{B}_0 \\ \widetilde{B}_0 & \widetilde{B}_1 \end{bmatrix} \begin{Bmatrix} M_0 \\ M_1 \end{Bmatrix} \quad \text{with the details}$$

$$\widetilde{B}_1 := \frac{1}{D}(\alpha \cosh \alpha \sin \beta - \beta \sinh \alpha \cos \beta)$$

$$\widetilde{B}_0 := \frac{1}{D}(\alpha \sin \beta - \beta \sinh \alpha)$$

$$D = (\alpha^2 + \beta^2) \sinh \alpha \sin \beta \tag{6.10}$$

Figure 6.2 shows for positive as well as for negative ϕ (actual for Winkler support) these generalized Berry functions. The singular behaviors are discussed in the following sections.

6.3 Eigenfrequency as a Function of Conservative Axial Load

As discussed earlier for the case of $\phi = 0$ the $\widetilde{B}$ functions are the basic Berry functions. For the case of $\lambda = 0$ then as seen from (6.7) $\alpha = \beta = \sqrt[4]{\phi}$ and then (6.10) with D integrated in the $\widetilde{B}$ functions is specifically

$$\widetilde{B}_1 = \frac{1}{2\sqrt[4]{\phi}}(\coth \sqrt[4]{\phi} - \cot \sqrt[4]{\phi}) \qquad \widetilde{B}_0 = \frac{1}{2\sqrt[4]{\phi}}(\frac{1}{\sinh \sqrt[4]{\phi}} - \frac{1}{\sin \sqrt[4]{\phi}})$$

$$D = 2\sqrt{\phi} \sinh \sqrt[4]{\phi} \sin \sqrt[4]{\phi} \tag{6.11}$$

and the singularity $D = 0$ gives the eigenvalue by $\sin \sqrt[4]{\phi} = 0$, i.e., $\sqrt[4]{\phi} = n\pi$ for $n = 1, 2, \ldots$. The $\tilde{}$ (tilde) functions in (6.10) are sometimes termed Koloušek frequency functions, see Koloušek (1973).

The eigenvalues for $D = 0$ in (6.8) are $\beta = n\pi$, this is inserted in (6.7) and squared to

$$(n\pi)^2 = -\frac{\lambda}{2} + \sqrt{\lambda^2/4 + \phi} \quad \rightarrow \quad \omega_n^2 = (n\pi)^2((n\pi)^2 + \lambda)\frac{EI}{L^4\rho A} \qquad (6.12)$$

For the lowest frequency equal to zero, the critical non-dimensional column force is determined $\lambda_C = -\pi^2$.

The eigenvalues without column force ($\lambda = 0$) is also seen directly from (6.12). For the present case of simply supports, the squared eigenfrequencies ω_n^2 are linear depending in the non-dimensional column force λ. For dependence with other BC, see Sect. 6.4 and Chap. 8.

6.4 Rotational Spring Supports

With rotational spring end supports as in Fig. 4.5, the end moments are

$$M_0 = -S_0\theta_0 = -S_0 y'(0)/L, \quad M_1 = -S_1\theta_1 = -S_1 y'(1)/L \qquad (6.13)$$

and two of the BC in (2.12) are with non-dimensional stiffnesses s_0, s_1 changed to

$$y''(0) - s_0 y'(0) = 0, \quad y''(1) + s_1 y'(1) = 0 \qquad (6.14)$$

The determinant D in (6.10) with these BC is more complicated, but still manageable and from a fourth order determinant found to

$$(\alpha^2 + \beta^2)^2 \sinh\alpha \sin\beta + (s_0 + s_1)(\alpha^2 + \beta^2)(\alpha\cosh\alpha\sin\beta - \beta\sinh\alpha\cos\beta) +$$

$$s_0 s_1(2\alpha\beta(1 - \cosh\alpha\cos\beta) + (\alpha^2 - \beta^2)\sinh\alpha\sin\beta) = 0 \qquad (6.15)$$

a special case of an even more general case in Chap. 8, i.e., (6.15) from (8.5), (8.6).

With non-dimensional stiffnesses s_0, s_1 and column force λ (6.14) may be solved iteratively by the Newton–Raphson method to obtain the results shown in Fig. 6.3. Note, the well-known results without column force ($\lambda = 0$) and the critical forces λ_C when the eigenfrequency is zero. The linear dependence in (6.12) is no longer valid, but is seen to be a practical approximation also for $s_0, s_1 \neq 0, 0$.

Beam-columns of cantilever types with translational end support are covered in Chap. 8, and it is then seen that a linear approximation cannot be applied for these cases.

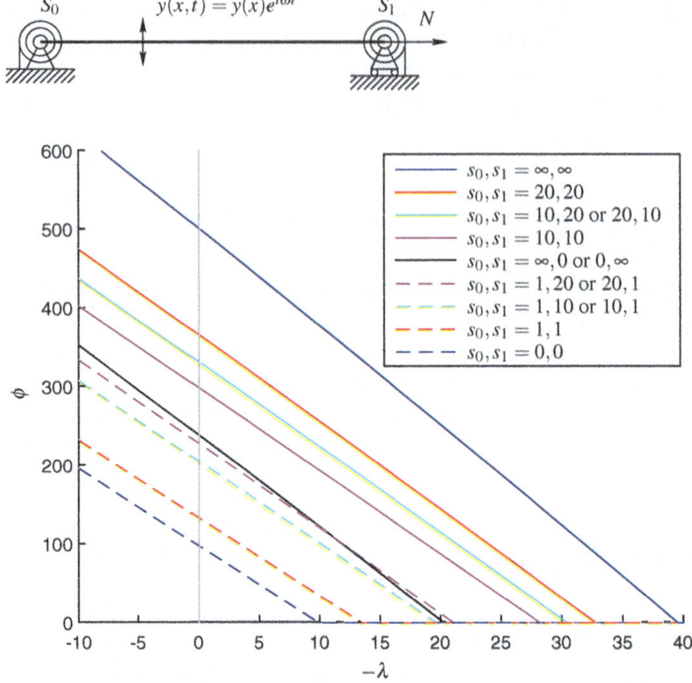

Fig. 6.3 Smallest (first) squared eigenfrequency $\omega^2 = \dfrac{\phi E I}{\rho A L^4}$ for beam-columns as function of non-dimensional column force λ, with end rotational spring stiffnesses as parameters

6.4.1 Euler Cases as Limiting Cases

Results for some classical cases (Euler cases) are obtained as special cases of (6.15). For a fixed—fixed beam ($s_0 = s_1 = \infty$) (6.15) gives

$$2\alpha\beta(1 - \cosh\alpha\cos\beta) + (\alpha^2 - \beta^2)\sinh\alpha\sin\beta = 0 \tag{6.16}$$

that without axial force ($\lambda = (\alpha^2 - \beta^2) = 0$) and ($\alpha = \beta = \sqrt[4]{\phi}$) gives

$$\cosh\sqrt[4]{\phi}\cos\sqrt[4]{\phi} = 1 \tag{6.17}$$

Another limiting case is fixed—simply supported BC that with $s_0 = 0, s_1 = \infty$ from (6.15) gives

$$\alpha\cosh\alpha\sin\beta - \beta\sinh\alpha\cos\beta = 0 \quad \rightarrow \quad \tan\sqrt[4]{\phi} = \tanh\sqrt[4]{\phi} \tag{6.18}$$

The characteristic equations (6.17) and (6.18) are derived differently in Chap. 3.

Chapter 7
Buckling with Spring Supported BC

This chapter is initiated with a discussion of the concept of stability, because it is not well defined. The name stability covers broadly, and it is not limited to the subject of the present book. There is no unique definition but different subjective, problem-oriented definitions. The American mathematician Bellman characterizes by the statement "Stability is a heavy loaded concept with an unstable definition."

Buckling with end rotations are solved by inverse evaluation, which is also possible for cases with right end translational spring supports. It is verified that the lowest number of critical buckles may be larger than one when Winkler supports are involved.

7.1 Mathematical Definition and Physical Experiments

Roughly stated in the present book, instability means too large displacements caused by small load changes. Figure 7.1a shows the result for displacement δ as a function of compressive column force P for a physical experiment, and it is seen that no well-defined instability force exists here. The result is without clear separation divided into three domains A, B, and C.

In domain A for relatively low compressive column force P we see small displacements and a normal beam response. In domain B the displacement δ increases fast, even with small increments of the compressive force, in agreement with the earlier shown Berry functions. Domain A and B are therefore computationally covered by the presented beam-column theory. The response in domain C is computationally more complicated, because many of the assumptions in Chap. 2 are violated. Later, chapters describe some extended formulations.

© Springer International Publishing AG, part of Springer Nature 2018
S. L. Wiggers and P. Pedersen, *Structural Stability and Vibration*, Springer Tracts
in Mechanical Engineering, https://doi.org/10.1007/978-3-319-72721-9_7

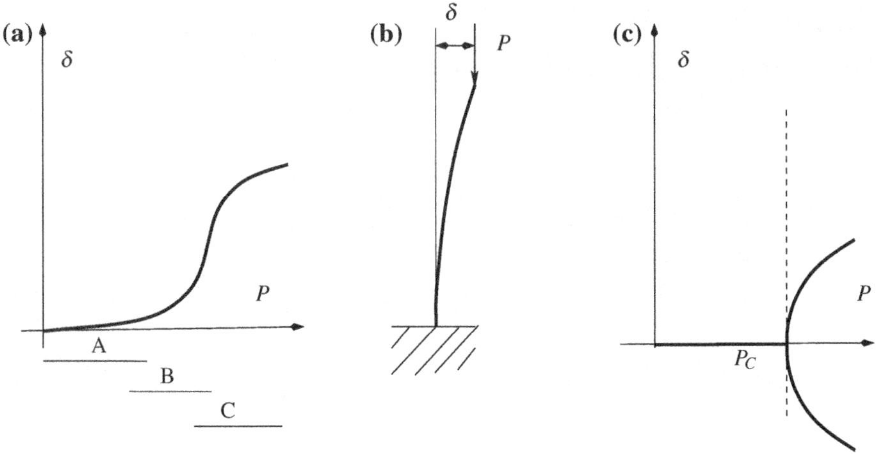

Fig. 7.1 Column model (**b**) with result of a physical experiment (**a**) and calculated results (**c**)

7.2   Different Instability Formulations

Figure 7.1c shows the result for an idealized column model. With this picture as reference, the different stability formulations (also named stability methods and stability criteria) are discussed.

The first method is named the method of equilibrium or the Euler method. It is analyzing an idealized model (homogeneous mathematical eigenvalue model), to find if a state of equilibrium infinitesimal close to and still different from the trivial state of equilibrium is possible, then the state is critical. This formulation results in $\delta = 0$ for $P < P_C$ as shown in Fig. 7.1c, and for $P = P_C$, the displacement δ is arbitrary (dotted line).

Method of equilibrium

The second method is named the energy method and leads to identical result as the first method. The analysis determines when the total potential energy is no longer positive definite, which is identical to the equilibrium method. The name energy method is often also used for an approximate solution of differential equations.

Energy method

The third method is named the imperfection method, like the application of beam-column theory and determine singularities. Imperfections may in principle be arbitrary: excentric loaded column force, nonlinear beamline, inhomogeneous material, or beam loads on columns. The critical column force is determined as the column force where displacements may be infinity large, independent of the size of the imperfections.

Imperfection method

The result for $P > P_C$ in Fig. 7.1c is later analytical exemplified in Chap. 12. The aim of the present chapter is to determine the critical column force, also named the buckling force for a beam-column.

Dynamic method

The methods described above are characterized by not including inertia forces, i.e., assuming static problems. It is therefore assumed that the stability is static or quasi-static (not influenced by inertia forces). It can be shown that this is the case for conservative problems. However, non-conservative problems that primarily are described as static may loose their stability dynamically. In Chap. 9 a dynamic formulation is as an introduction applied for such a problem. For this method, dynamic equilibrium is formulated and instability need not be with zero frequency (divergent instability), but flutter (dynamic instability) may occur.

7.3 Buckling with End Rotations

The beam-columns analyzed in Chaps. 2 and 4 may buckle with end rotations. The functions shown in Figs. 4.2, 4.4, 4.7, 5.2 and 5.3 show that end rotations can be large even for infinitesimal beam loads M, Q, or q. It is seen that although a specific beam-column result is depending on the class of beam load (M or Q) as well as on the position of the load η, then the column force N_C that result in a singular behavior only depends on the BC for the beam-column. This column force is named the buckling force or the critical column force.

Simply supported

For simply supported beam-columns, the Berry functions B_0 and B_1 defined in (4.15) show in Fig. 4.2 singularities for these functions if

$$\sin \sqrt{-\lambda_C} = 0 \quad \rightarrow \quad \sqrt{-\lambda_C} = n\pi \quad \text{for} \quad n = 0, 1, 2, \ldots \tag{7.1}$$

The case of $n = 0$ corresponds to a trivial solution and a dynamic formulation of column equilibrium shows that in almost all cases only $n = 1$ is of practical interest, giving the buckling force.

$$N_C = -\pi^2 \frac{EI}{L^2} \tag{7.2}$$

as derived analytical in Chap. 3. For a case, where $n > 1$ is actual, see Sect. 7.5.

Rotational springs

The buckling force is depending on the actual BC. Therefore, it is important to determine the buckling force for the beam-columns in Sect. 4.3. The more general beam-column solution is presented by (4.19), and a singularity exists for the characteristic equation $D = 0$. In terms of the column force (by the Berry functions B_0, B_1) and the rotational non-dimensional stiffnesses s_0, s_1, a singularity exists for the characteristic equation $D = 0$.

$$D = 1 + (s_0 + s_1)B_1 + s_0 s_1 (B_1^2 - B_0^2) = 0 \tag{7.3}$$

Inverse evaluation

For given stiffnesses s_0, s_1, this is a nonlinear equation to determine λ_C, say by the Newton–Raphson iterations. However, the inverse problem (given λ_C and say s_1 find s_0) is solved explicitly by

$$(s_0)_C = \frac{1 + s_1 B_1}{B_1 + s_1 (B_1^2 - B_0^2)} \tag{7.4}$$

Figure 7.2 shows evaluated solutions with s_1 as parameter. Note, especially in Fig. 7.2, the high sensitivity of λ_C at small values of s_0 close to $s_0 = 0$.

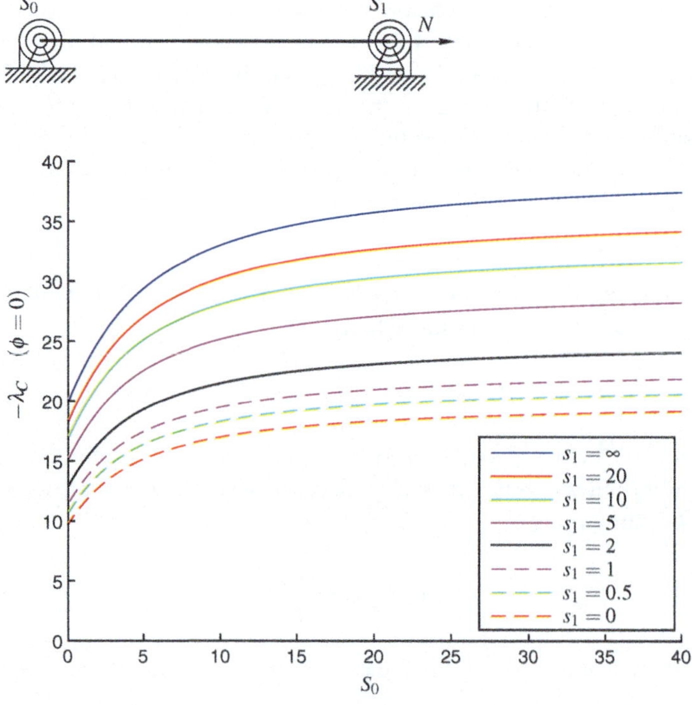

Fig. 7.2 Buckling column force $N_C = \frac{\lambda_C EI}{L^2}$ for columns with end rotational spring support

Limiting cases

Three cases in Fig. 7.2 show agreement with earlier presented results for Euler cases

- simply—simply supported by $s_0 = s_1 = 0 \;\rightarrow\; \lambda_C = -\pi^2$
- fixed—simply supported by $s_0 = \infty, s_1 = 0 \;\rightarrow\; \lambda_C \simeq -20.2$
- simply—fixed supported by $s_0 = 0, s_1 = \infty \;\rightarrow\; \lambda_C \simeq -20.2$
- fixed—fixed supported by $s_0 = s_1 = \infty \;\rightarrow\; \lambda_C = -4\pi^2$

Note, the relations between the buckling forces that are 1 to almost 2 to 4 for these three Euler cases.

7.4   Buckling with End Translations

As it follows from the Euler case III of fixed—free supports, the buckling is especially critical for cantilever beam-column types. This is due to the larger beam moments resulting from the translated position of the column force.

Cantilever buckling

For problems more complicated than the Euler case III and including translational stiffness at the right end, the general beam-column solution is presented in (5.14) and (5.15) with singular behavior for $D = 0$. Large displacements result for even infinitesimal beam loads if

$$(k + \lambda)\left(1 + (s_1 + s_0)B_1 + s_1 s_0(B_1^2 - B_0^2)\right) + 2s_1 s_0(B_1 - B_0) + (s_1 + s_0) = 0 \tag{7.5}$$

Inverse evaluation

As applied before, this equation is solved explicitly by the inverse problem, i.e., given $\lambda_C \rightarrow B_0, B_1$ and s_0, s_1, the critical non-dimensional translational force k_C is evaluated.

$$k_C = -\frac{2s_1 s_0(B_1 - B_0) + s_1 + s_0}{1 + (s_1 + s_0)B_1 + s_1 s_0(B_1^2 - B_0^2)} - \lambda \tag{7.6}$$

Figures 7.3, 7.4, and 7.5 show the evaluated results. The rotational spring s_1 is chosen as parameter, and the three figures assume the other rotational springs $s_0 = \infty, 1$, and 0. The case of $s_0 = \infty$ is of specific interest, simplifying (7.6) to

$$k_C = -\frac{2s_1(B_1 - B_0) + 1}{B_1 + s_1(B_1^2 - B_0^2)} - \lambda \tag{7.7}$$

with evaluated solutions presented in Fig. 7.3.

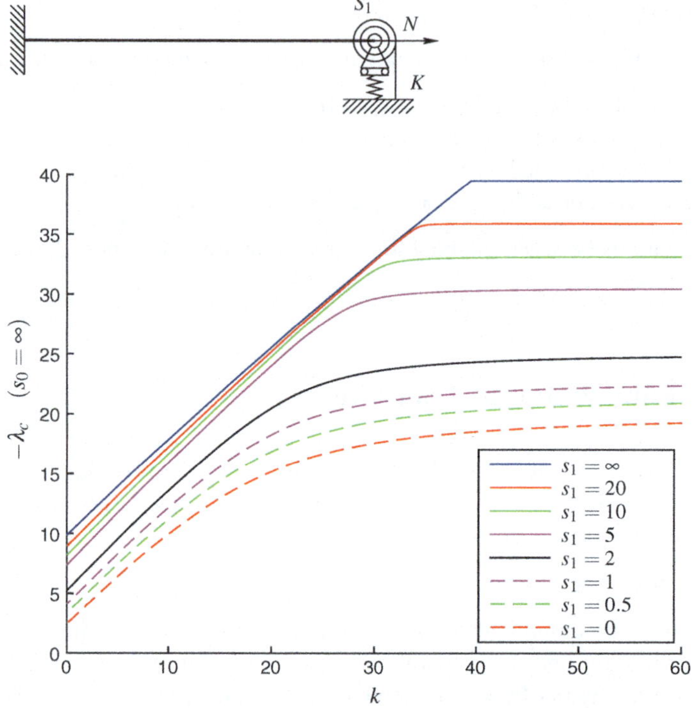

Fig. 7.3 Buckling force $N_C = \frac{\lambda_C EI}{L^2}$ for beam-columns of cantilever type, as a function of stiffness for translational spring, with non-dimensional rotational end spring stiffness s_1 as parameter and $s_0 = \infty$

Another limiting case corresponding to a free end by $k = s_1 = 0$, giving

$$(s_0)_C = -\frac{\lambda}{\lambda B_1 + 1} \tag{7.8}$$

Extending this case to include $k \neq 0$, but still $s_1 = 0$ gives

$$(s_0)_C = -\frac{k + \lambda}{(k + \lambda) B_1 + 1} \tag{7.9}$$

with evaluated results in Fig. 7.6. Figure 7.6 presents results differently, now with k as parameter.

Limiting Euler cases

Cases in Figs. 7.3 and 7.4 show agreement with earlier presented results for Euler cases

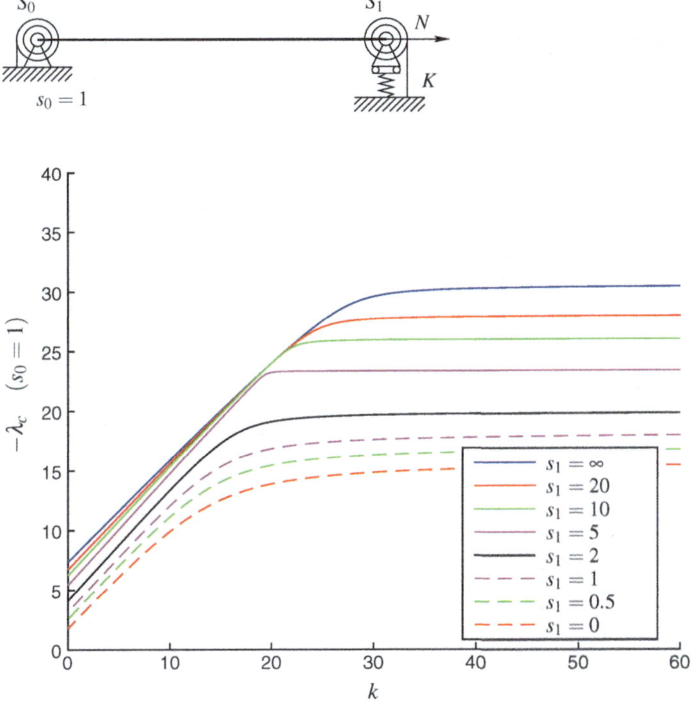

Fig. 7.4 Buckling force $N_C = \frac{\lambda_C EI}{L^2}$ for beam-columns of cantilever type, as a function of stiff-nesses for end springs, here specifically $s_0 = 1$

- fixed—free supported by $s_0 = \infty, s_1 = k = 0 \;\rightarrow\; \lambda_C = -\frac{\pi^2}{4}$
- fixed—translational sliding supported by $s_0 = s_1 = \infty, k = 0 \;\rightarrow\; \lambda_C = -\pi^2$

Note, the relation between the buckling forces that is 1/4 to 1 for these two Euler cases.

7.5 Buckling with Winkler Support

Beam-columns may be continuously supported by linear elastic springs (Winkler support) as shortly described in Chap. 6. Figure 7.7 shows the model to be exemplified.

The non-dimensional parameter ϕ may from (2.8) correspond to

$$\phi y = \frac{L^4 q}{EI} \tag{7.10}$$

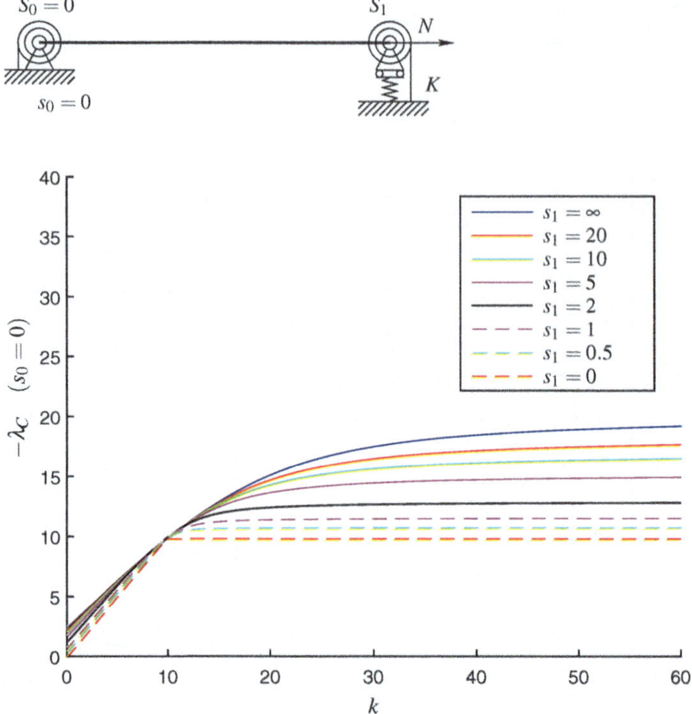

Fig. 7.5 Buckling force $N_C = \frac{\lambda_C EI}{L^2}$ for beam-columns of cantilever type, as a function of stiffnesses for end springs, here specifically $s_0 = 0$

that with both inertia forces per length $q = \omega^2 \rho A y$ in (2.11) and now in addition from reaction force (per length proportional to and against displacement y) is $q = -ry$, together give

$$\phi = \omega^2 \frac{L^4 \rho A}{EI} - r \frac{L^4}{EI} \quad \rightarrow \quad \phi \frac{EI}{L^4} = \omega^2 \rho A - r \quad \text{or}$$

$$\omega^2 = \frac{1}{\rho A}\left(\phi \frac{EI}{L^4} + r\right) \tag{7.11}$$

showing that the squared eigenfrequency is linear increasing as a function of support stiffness parameter r. The solution for simply—simply end supports is determined in Chap. 6 to

$$\sin \beta = 0 \quad \rightarrow \quad \beta = n\pi \quad \text{for} \quad n = 1, 2, \dots \text{ with}$$

$$\beta = \sqrt{-\lambda/2 + \sqrt{\lambda^2/4 + \phi}} \tag{7.12}$$

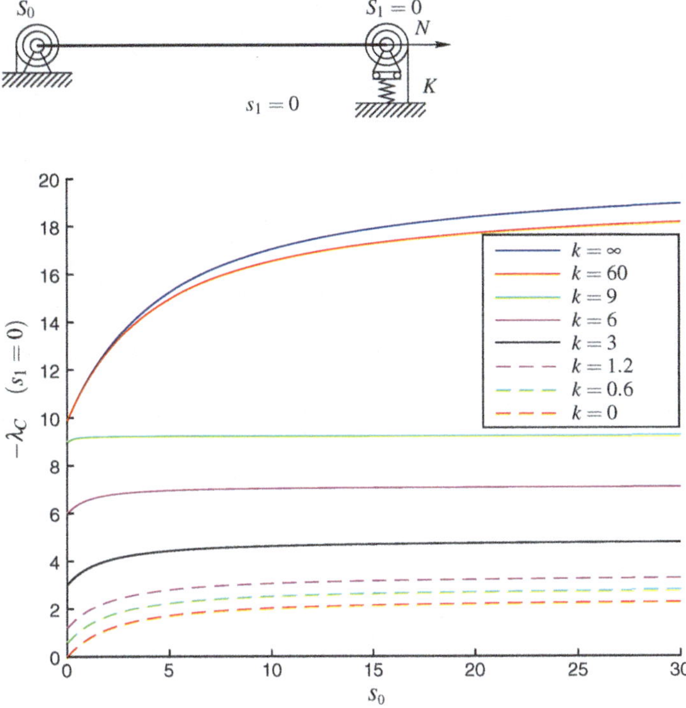

Fig. 7.6 Buckling force $N_C = \frac{\lambda_C EI}{L^2}$ for beam-columns of cantilever type, as a function of stiffnesses for end springs, here specifically $s_1 = 0$ and k as a parameter

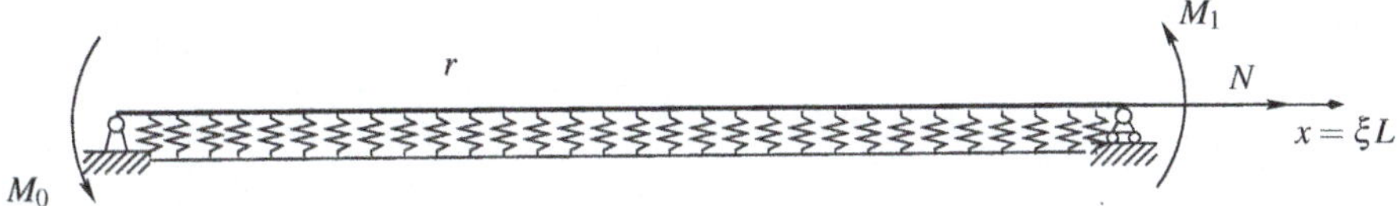

Fig. 7.7 Beam-column continuously supported by linear elastic springs (Winkler support) with spring stiffness r

Critical number of buckles

The critical column force is the lowest solution to (7.12), and for the case with continuously support r, the lowest value of λ_C is not always corresponding to $n = 1$. For the pure static problem with $\phi = \frac{-r L^4}{EI}$, this follows from

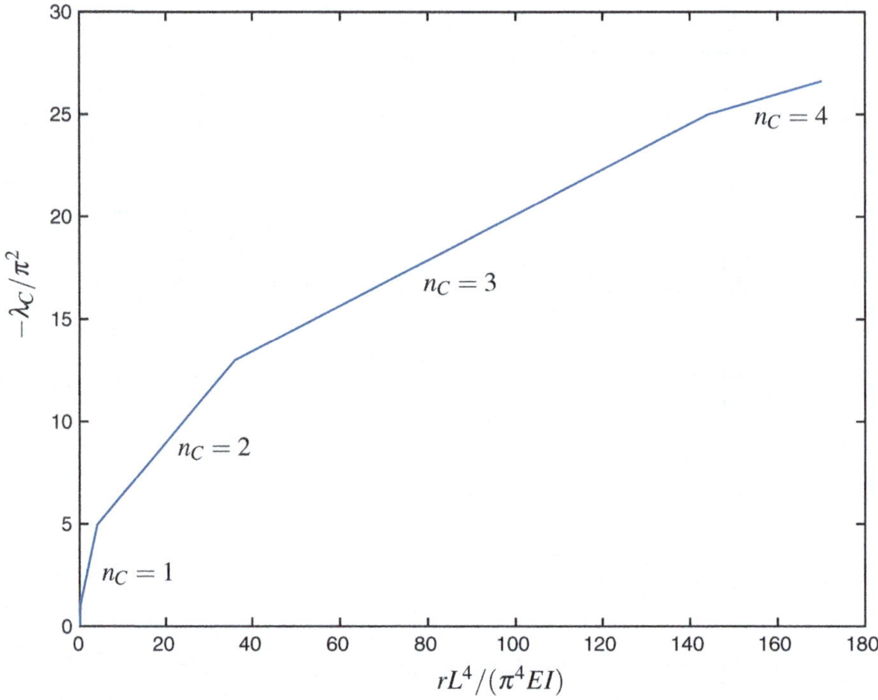

Fig. 7.8 Number of buckles for critical column force of a beam-column continuously supported by linear elastic springs (Winkler support), as a function of the relative spring stiffness $rL^4/(\pi^4 EI)$

$$(n\pi)^2 = -\frac{\lambda_C}{2} + \sqrt{\frac{\lambda_C^2}{4} - \frac{rL^4}{EI}} \rightarrow \left((n\pi)^2 + \frac{\lambda_C}{2}\right)^2 = \frac{\lambda_C^2}{4} - \frac{rL^4}{EI}$$

$$\rightarrow (n\pi)^4 + \lambda_C(n\pi)^2 = -\frac{rL^4}{EI} \rightarrow \lambda_C = -n^2\pi^2 - \frac{1}{n^2\pi^2}\frac{rL^4}{EI} \tag{7.13}$$

Note that the first term in the final result (7.13) is numerically lowest for $n = 1$ while the second term is numerically lowest for $n \rightarrow \infty$. The critical number of n is therefore depending on the relation $\frac{rL^4}{EI}$, i.e., the ratio between the stiffness rL^4 and the bending stiffness EI of the beam. The lowest critical column force λ_C corresponds to the lowest elastic energy (necessary for buckling) in the total system with also energy in the distributed supports. Figure 7.8 shows λ_C as a function of $\frac{rL^4}{EI}$, indicating the critical number of buckles n_C.

7.5.1 Eigenfrequencies with Winkler Support

Combining vibration with Winkler support, the new definition of ϕ is

$$\phi = (\omega^2 \rho A - r)\frac{L^4}{EI} \tag{7.14}$$

Assuming that the squared eigenfrequency without Winkler support is ω_0^2, then the modified squared eigenfrequency with Winkler support (unchanged ϕ) is

$$\omega^2 \rho A - r = \omega_0^2 \rho A \quad \rightarrow \quad \omega^2 = \omega_0^2 + \frac{r}{\rho A} \tag{7.15}$$

Magnetic attraction

Note that including simple linear model for magnetic attraction described by factor a, the results are unchanged with the substitution $r \rightarrow (r - a)$.

Chapter 8
Eigenfrequencies of Beam-Columns with Spring Supported BC

The general eigenvalue problem with three supporting parameters (non-dimensional s_0, s_1, and k) and also combined axial load and eigenfrequency is described with its general solution. This includes the specific cases treated in the previous chapters, and the general case is mathematically more demanding. Characteristic equations for all cases in Fig. 2.2 are derived. These transcendental equations may for some cases be solved to explicit solutions, using an inverse approach. Other presented results are obtained applying the Newton–Raphson method. To illustrate the limitations by assumption of conservative external axial force, the lowest eigenfrequency as a function of a non-conservative follower column force is presented. Next chapter gives an introduction to such dynamic stability formulation.

The vibration of beam-columns is the most simple example of the integrated treatment of stability and vibration. The results of the present chapter are available in detail in (Pedersen 1986) and include linear elastic supports, rotational as well as translational. The focus is on squared eigenfrequencies as a function of column force and elastic support.

8.1 Eigenvalue Problems with Analytical Solution

The general DE (2.12) in Chap. 2 is repeated

$$y'''' - \lambda y'' - \phi y \equiv 0 \tag{8.1}$$

with the two non-dimensional constants for column force λ and for squared frequency ϕ by

© Springer International Publishing AG, part of Springer Nature 2018
S. L. Wiggers and P. Pedersen, *Structural Stability and Vibration*, Springer Tracts
in Mechanical Engineering, https://doi.org/10.1007/978-3-319-72721-9_8

Non-dimensional constants

$$\lambda := \frac{NL^2}{EI} \quad \text{and} \quad \phi := \frac{\omega^2 L^4 \rho A}{EI} \tag{8.2}$$

The general BC treated in the present chapter is

$$y(0) = 0, \quad y''(0) - s_0 y'(0) = 0,$$
$$y''(1) + s_1 y'(1) = 0, \quad y'''(1) - (1 - \gamma)\lambda y'(1) - ky(1) = 0 \tag{8.3}$$

As in (6.6) of Chap. 6, the general solution to (8.1) is

$$y(\xi) = C_1 \cosh(\alpha\xi) + C_2 \sinh(\alpha\xi) + C_3 \cos(\beta\xi) + C_4 \sin(\beta\xi) \quad \text{with}$$

$$\alpha := \sqrt{\frac{\lambda}{2} + \sqrt{\frac{\lambda^2}{4} + \phi}} \quad \text{and} \quad \beta := \sqrt{-\frac{\lambda}{2} + \sqrt{\frac{\lambda^2}{4} + \phi}}$$

$$\lambda = \alpha^2 - \beta^2, \quad \phi = \alpha^2\beta^2 \quad \text{and} \quad \alpha^2 + \beta^2 = 2\sqrt{\frac{\lambda^2}{4} + \phi} \tag{8.4}$$

The case of $\gamma \neq 0$ corresponds to a non-conservative column force and will first be discussed in detail in Chap. 9.

Total characteristic equation

Inserting the solution (8.4) with derivatives as specified in (6.6) in the four BC and evaluating the fourth-order determinant D that gives the characteristic equation by $D = 0$ (this is done with help of an algebraically computer language, like Mathematica (Wolfram 1991), the characteristic equation is written

$$D = k(D_1 + (s_0 + s_1)D_2 + s_0 s_1 D_3) + (1 - \gamma)\lambda(D_4 + s_0 D_5)$$
$$+ (\sqrt{-\phi})^2 D_2 + s_0 D_6 + s_1 D_7 + s_0 s_1 D_4 = 0 \tag{8.5}$$

with the seven defined sub-functions D_i expressed solely in terms of the quantities α and β that are defined in (8.4) by λ and ϕ

$$D_1 := (\alpha^2 + \beta^2)^2 \sinh\alpha \sin\beta$$
$$D_2 := (\alpha^2 + \beta^2)(\alpha \cosh\alpha \sin\beta - \beta \sinh\alpha \cos\beta)$$
$$D_3 := 2\alpha\beta(1 - \cosh\alpha \cos\beta) + (\alpha^2 - \beta^2)\sinh\alpha \sin\beta$$
$$D_4 := \alpha\beta(\alpha^2 + \beta^2)(\beta \cosh\alpha \sin\beta + \alpha \sinh\alpha \cos\beta)$$

$$D_5 := \alpha\beta(2\alpha\beta \sinh\alpha \sin\beta) - (\alpha^2 - \beta^2)(1 - \cosh\alpha \cos\beta)$$

$$D_6 := \alpha\beta((\alpha^2 + \beta^2)^2 - \alpha\beta D_3)$$

$$D_7 := \alpha\beta(\alpha^2 + \beta^2)^2 \cosh\alpha \cos\beta \tag{8.6}$$

Conservative column force

For a conservative problem, in (8.5) $\gamma = 0$, the goal is to determine

$$\phi = \phi(s_0, s_1, k, \lambda) \quad \rightarrow \quad \omega^2 = \phi\frac{EI}{\rho AL^4} \tag{8.7}$$

by Newton–Raphson solutions to the transcendental Eq. (8.5) and then illustrate the solutions graphically, which is here only done for the lowest squared eigenfrequency.

Cases of $k = \infty$

Firstly, (8.5) is interpreted for the more simple cases treated in the earlier chapters. The case of no end translations corresponds to $k = \infty$, and thus, only the first part of (8.5) is involved in this case, as shown in Fig. 8.1a

$$D_1 + (s_0 + s_1)D_2 + s_0 s_1 D_3 = 0 \tag{8.8}$$

From (8.8), case (b) in Fig. 8.1 is obtained by omitting terms with $s_0 = 0$ and further case (c) from case (b) omitting terms with $s_1 = 0$, i.e., the Euler I case is obtained. Case (d) follows from case (a) when only terms proportional to s_0 are included and further case (e) by omitting terms with s_1, i.e., the Euler V case is obtained. Case (f) only includes terms proportional to $s_0 s_1$, i.e., the Euler II case is obtained.

Cases of $k \neq \infty$

From the general case (g) presented in (8.5) case (h) only includes terms proportional to s_0. From case (h), omitting terms proportional to s_1 gives the result for case (i). The BC for case (j) corresponds to $s_1 = \infty$, and therefore, only terms in (h) proportional to s_1 are included. Finally, from case (j) with $k = 0$ case (k) is obtained, i.e., the Euler IV case is obtained. From case (i) with $k = 0$ case (l), i.e., the Euler III case, is obtained.

BC-specific characteristic equations

Fig. 8.1 shows the specific characteristic equations for all the studied BC, initially presented in Chap. 2. The resulting characteristic equations for $\lambda = 0$ corresponding to the characteristic equations for the five Euler cases are for case I $D_1 = 0$, for case

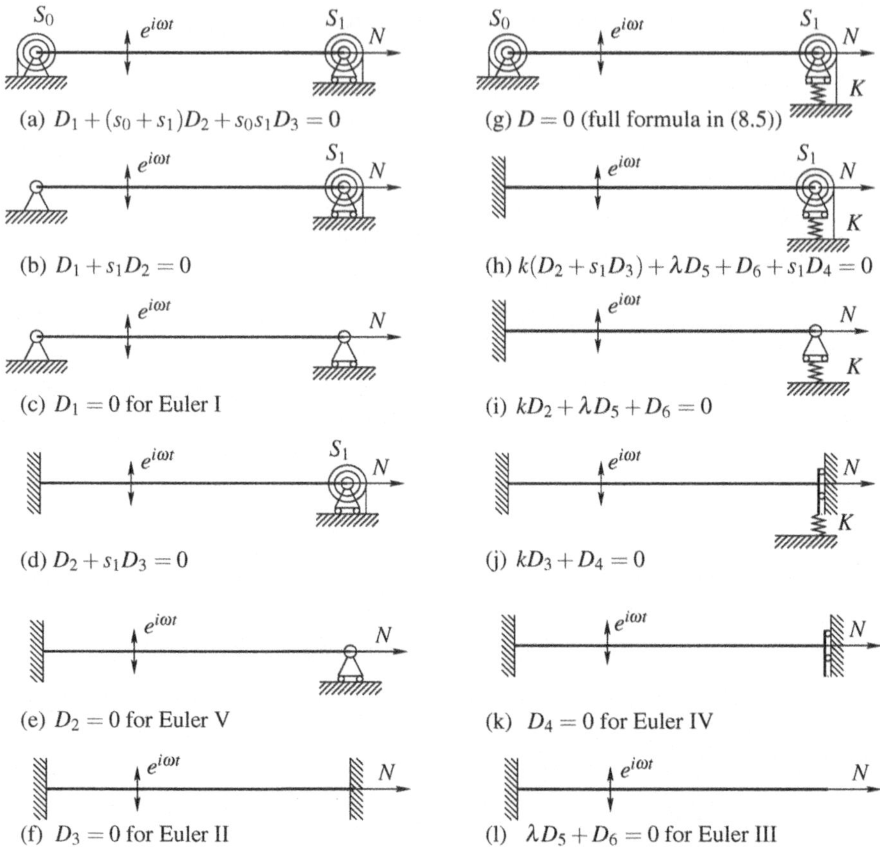

Fig. 8.1 Characteristic equations (from (8.5) for $\gamma = 0$), corresponding to the boundary conditions (BC) of the studied beam-column cases

II $D_3 = 0$, for case III $D_6 = 0$, for case IV $D_4 = 0$, and for case V $D_2 = 0$. This can be seen to agree with the results in Table 3.2 in Chap. 3.

8.2 Solving the Transcendental Equations

The seven sub-functions D_i are only functions of λ and ϕ, here written in terms of α, β. A number of questions may be answered by solving the characteristic equations that are linear combinations of these functions with the values for the BC, i.e., with s_0, s_1, k as parameters.

- Without column force ($\lambda = 0$), how do the squared eigenfrequencies depend on the non-dimensional boundary stiffnesses s_0, s_1, k?

- With column force ($\lambda \neq 0$) in tension as well as in compression, how do the squared eigenfrequencies depend on λ treating s_0, s_1, k as parameters?

Newton–Raphson iterative method

The presented numerical solutions correspond only to the lowest squared eigenfrequency, but similar results for higher-order squared eigenfrequencies may be obtained by the applied procedure, using different starting conditions for the Newton–Raphson method. The iterative approach of Newton–Raphson is shortly described in Chap. 16.

8.3 Explicit Solutions by Inverse Approach

Explicit solutions may be applied for pure stability problems ($\phi = 0$) or for pure vibration problems ($\lambda = 0$). For the pure stability problems, numerical solutions are presented in Figs. 7.2, 7.3, 7.4, 7.5 and 7.6. For these cases, the sub-functions D_i can alternatively be written in terms of the Berry functions.

For the pure vibration problem, it follows from (8.4) that

$$\alpha = \beta = \sqrt[4]{\phi} \;\Rightarrow\; \alpha^2 - \beta^2 = 0 \text{ and } \alpha^2 + \beta^2 = 2\sqrt{\phi} \tag{8.9}$$

and the seven sub-functions are then

$$
\begin{aligned}
D_1 &= 4\phi \sinh\sqrt[4]{\phi}\sin\sqrt[4]{\phi} \\
D_2 &= 2\sqrt[4]{\phi^3}(\cosh\sqrt[4]{\phi}\sin\sqrt[4]{\phi} - \sinh\sqrt[4]{\phi}\cos\sqrt[4]{\phi}) \\
D_3 &= 2\sqrt{\phi}(1 - \cosh\sqrt[4]{\phi}\cos\sqrt[4]{\phi}) \\
D_4 &= 2\phi\sqrt[4]{\phi}(\cosh\sqrt[4]{\phi}\sin\sqrt[4]{\phi} + \sinh\sqrt[4]{\phi}\cos\sqrt[4]{\phi}) \\
D_5 &= 2\phi \sinh\sqrt[4]{\phi}\sin\sqrt[4]{\phi} \\
D_6 &= \phi(4\sqrt{\phi} - D_3) \\
D_7 &= 4\phi\sqrt{\phi}\cosh\sqrt[4]{\phi}\cos\sqrt[4]{\phi}
\end{aligned}
\tag{8.10}
$$

Inverse explicit solutions

The inverse approach of treating a single BC parameter as unknown may result in an explicit solution, as exemplified for s_1 when $k = \infty$ and for k when $s_0 = \infty$.

$$s_1 = \frac{-D_1 - s_0 D_2}{D_2 + s_0 D_3} \text{ or } s_0 = \frac{-D_1 - s_1 D_2}{D_2 + s_1 D_3} \quad \text{for case a)}$$

$$s_1 = \frac{-D_1}{D_2} \quad \text{for case b)}$$

$$s_1 = \frac{-D_2}{D_3} \quad \text{for case d)}$$

$$k = \frac{-(\lambda D_5 + D_6 + s_1 D_4)}{D_2 + s_1 D_3} \quad \text{for case h)}$$

$$k = \frac{-\lambda D_5 - D_6}{D_2} \quad \text{for case i)}$$

$$k = \frac{-D_4}{D_3} \quad \text{for case j)} \tag{8.11}$$

where $\lambda = 0$ if there is no column force.

Such explicit solutions are applied and presented in Figs. 8.2 and 8.3. These two cases (a) and (h), without column force, are only presenting lowest squared eigenfrequency as a function of stiffnesses for BC.

8.3.1   Lowest Eigenfrequency as a Function of Support Stiffnesses, Assumed No Column Force

Eigenfrequencies for beams with elastic boundary supports are generally not available. This holds even for beams without column force ($\lambda = 0$), and such results are initially exemplified in Figs. 8.2 and 8.3. With S_0, S_1 as parameters, the case in Fig. 8.2 as in Fig. 6.3 solved also with column force ($\lambda \neq 0$), then applying the Newton–Raphson method.

8.4   Alternative Function Expressions

In general, the parameters α and β of (8.6) are complex quantities and the variable transformation

$$g := \beta + i\alpha \;\; \rightarrow \;\; g^2 = -\lambda + 2\sqrt{-\phi}$$

$$\bar{g} := \beta - i\alpha \;\; \rightarrow \;\; \bar{g}^2 = -\lambda - 2\sqrt{-\phi},$$

$$g\bar{g} = \alpha^2, \quad \alpha = -i\frac{1}{2}(g - \bar{g}), \quad \beta = \frac{1}{2}(g + \bar{g}) \tag{8.12}$$

which has previously, see (Pedersen 1977a), (Wittrick 1982) and (Morgan and Sinha 1983), been used successfully, gives more manageable functions, especially when these functions have to be differentiated with respect to λ and/or ϕ.

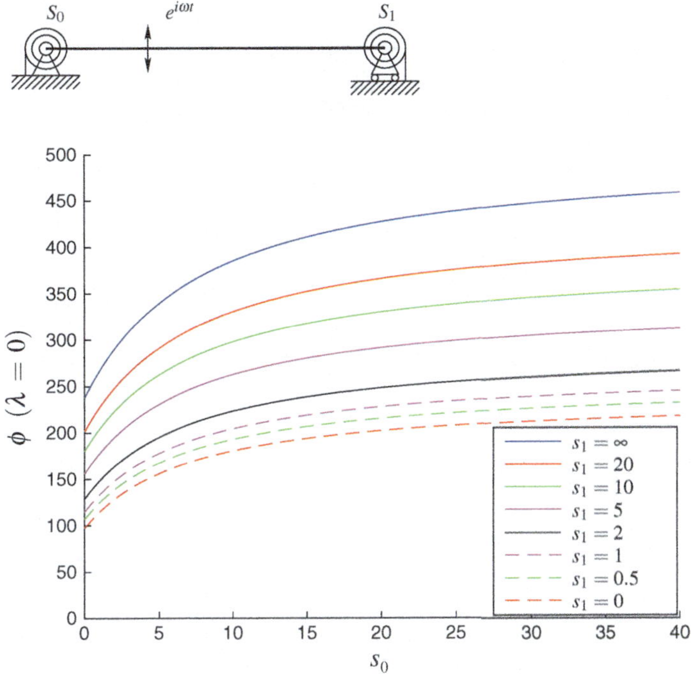

Fig. 8.2 Lowest eigenfrequency as a function of support stiffnesses, assumed no column force, only rotational end springs. Results obtained by inverse solutions to (8.11a)

Trigonometric summation formulas give

$$\cosh\alpha\cos\beta = (\cos g + \cos\bar{g}), \quad \cosh\alpha\sin\beta = (\sin g + \sin\bar{g}),$$
$$\sinh\alpha\cos\beta = (\sin g - \sin\bar{g}), \quad \sinh\alpha\sin\beta = (\cos g - \cos\bar{g}), \tag{8.13}$$

and then (upon omitting the common factor i), the seven D sub-functions are expressed as

$$D_1 := (\lambda^2 - 4(\sqrt{-\phi})^2)(\cos g - \cos\bar{g})$$
$$D_2 := g\bar{g}(\bar{g}\sin g - g\sin\bar{g})$$
$$D_3 := 2\sqrt{-\phi}(\cos g + \cos\bar{g} - 2) + \lambda(\cos g - \cos\bar{g})$$
$$D_4 := -\sqrt{-\phi}\,g\bar{g}(\bar{g}\sin g + g\sin\bar{g})$$
$$D_5 := -\sqrt{-\phi}\left(2\sqrt{-\phi}(\cos g - \cos\bar{g}) + \lambda(\cos g + \cos\bar{g} - 2)\right)$$

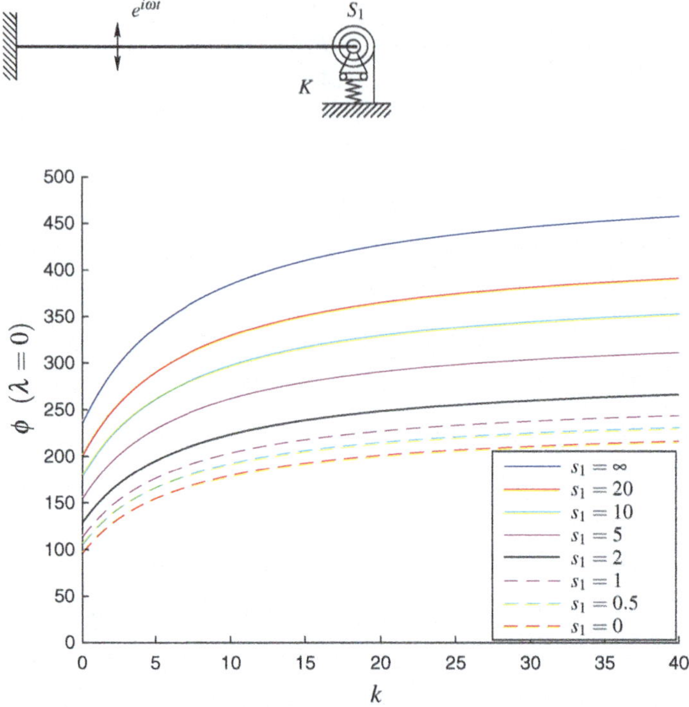

Fig. 8.3 Lowest eigenfrequency as a function of support stiffnesses, assumed no column force, also translational end spring. Results obtained by inverse solutions to (8.11h)

$$D_6 := -\sqrt{-\phi}\left(2(\lambda^2 - 4(\sqrt{-\phi})^2) - \sqrt{-\phi}D_3\right)$$

$$D_7 := -\sqrt{-\phi}(\lambda^2 - 4(\sqrt{-\phi})^2)(\cos g + \cos \bar{g}) \tag{8.14}$$

The sub-functions are not affected by a change in sign for either g or $\bar{g}$. One may thus choose $g = \sqrt{g^2}$, $\bar{g} = \sqrt{\bar{g}^2}$. A change in sign for $\sqrt{-\phi}$ interchanges g and $\bar{g}$ and thus causes a general sign change for all sub-functions. This does not influence (8.5), and one may again choose the positive square root.

A small computer program written with complex variables suffices to solve the general Eq. (8.5).

8.5 Specific Graphically Presented Results, Obtained by the Newton–Raphson Method

A number of specific solutions to (8.5) are shown with reference to a uniform Bernoulli–Euler modeled beam of length L, cross-sectional area A, bending stiffness EI, mass density ρ, and absolute values of boundary stiffnesses $S_0 = s_0 EI/L$,

$S_1 = s_1 EI/L$, and $K = kEI/L^3$. Then, ϕ gives the squared eigenfrequency from $\omega^2 = \phi EI/(\rho AL^4)$, and λ gives the column force N positive in "tension" from $N = \lambda EI/L^2$.

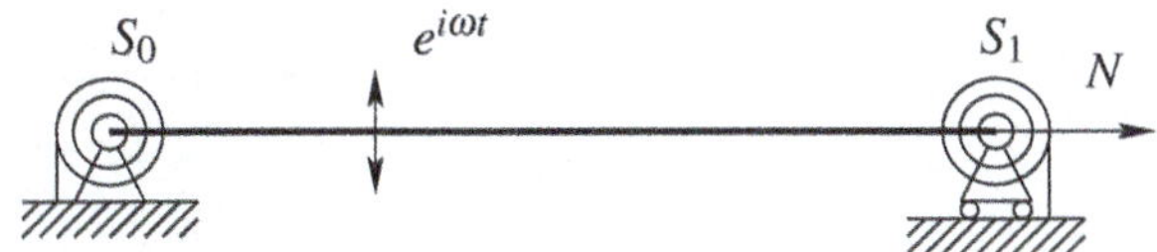

Fig. 8.4 Lowest eigenfrequency as a function of column force, with support stiffnesses as parameters, only rotational end springs. Results for this case are already shown in Fig. 6.3 obtained by applying the Newton–Raphson method

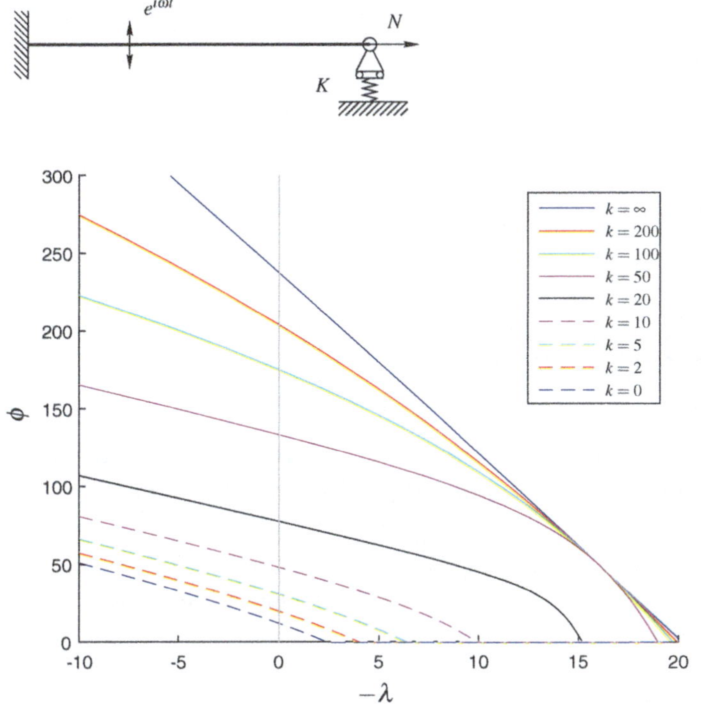

Fig. 8.5 Lowest eigenfrequency as a function of column force, with support stiffnesses as parameters, also translational end spring but left end fixed. Results obtained by applying the Newton–Raphson method to solve the characteristic equation in Fig. 8.1i

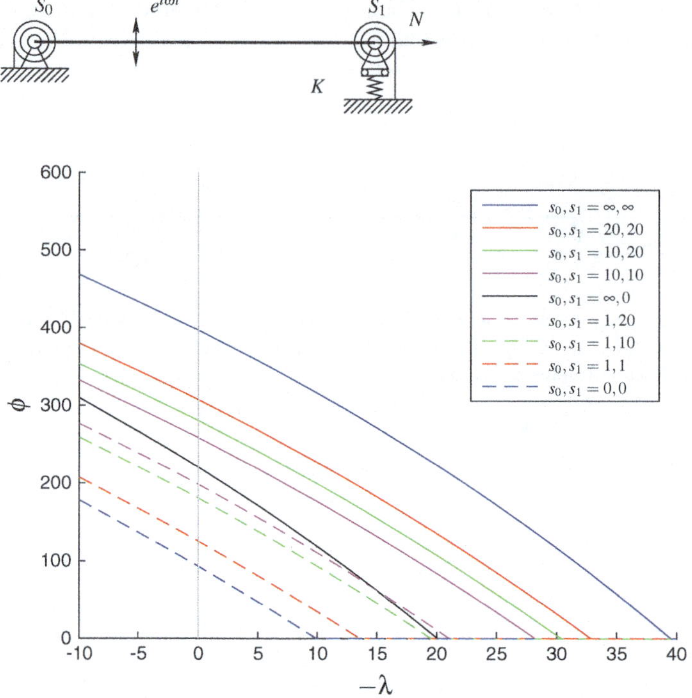

Fig. 8.6 Lowest eigenfrequency as a function of column force. The results are obtained by Newton–Raphson iterations for the general case in Fig. 8.1g, here with $k = 400$

8.5.1   Lowest Eigenfrequency as a Function of Column Force, with Support Stiffnesses as Parameters

The primary aim of the present chapter is to determine eigenfrequencies for beam-columns with column force λ and graphically illustrate the influence of this column force (Figs. 8.4 and 8.5).

8.5.2   Lowest Eigenfrequency as a Function of Column Force, Further BC Parameters

Figure 8.6 shows results for $k = 400$ with further parameters $(s_0, s_1) = (\infty, \infty)$, (10, 10), $(\infty, 0)$, (1, 10), and (0, 0) for the lowest eigenfrequency as a function of column force. The results are obtained by Newton–Raphson iterations for the general case in Fig. 8.1g.

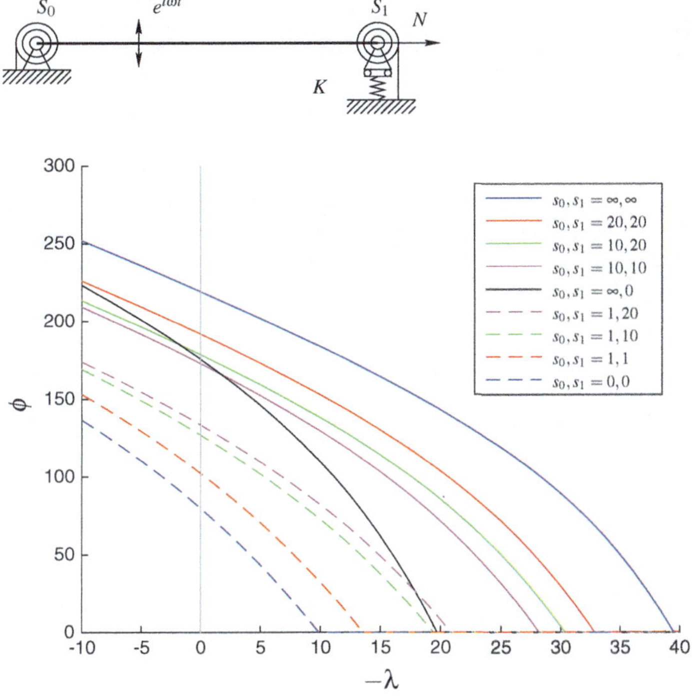

Fig. 8.7   Lowest eigenfrequency as a function of column force. The results are obtained by Newton–Raphson iterations for the general case in Fig. 8.1g, here with $k = 100$

Figure 8.7 shows results for a more flexible translational case of $k = 100$, but else identical to the assumptions behind Fig. 8.6

Figure 8.8 shows results for fixed left boundary ($s_0 = \infty$) as well as rotational rigid right boundary ($s_1 = \infty$). This corresponds to the case in Fig. 8.1j. The translational stiffness k is treated as parameter. The results are obtained by Newton–Raphson iterations for the case in Fig. 8.1j, i.e., by solving for $kD_3 + D_4 = 0$.

Figure 8.9 shows results for a more flexible rotational case of $s_1 = 2$, but else identical to the assumptions behind Fig. 8.8.

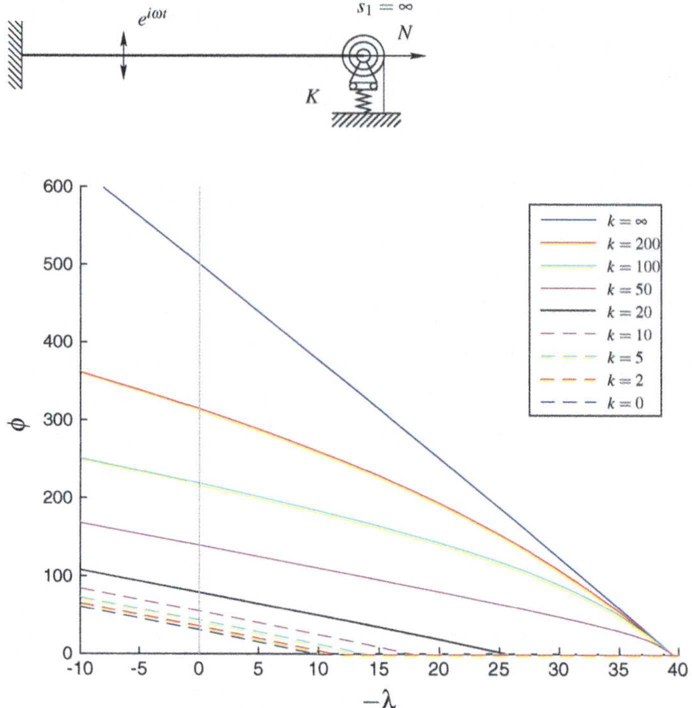

Fig. 8.8 Lowest eigenfrequency as a function of column force. The results are obtained by Newton–Raphson iterations for the case in Fig. 8.1j, here with no rotation at the right end $s_1 = \infty$

8.6   Lowest Eigenfrequency as a Function of Non-conservative "follower" Column Force

All results shown until now correspond to $\gamma = 0$, for which the eigenvalue problem of (8.1) with (8.3) is self-adjoint. Note from the presented results that ϕ and then ω^2 always decrease with increasing $|-\lambda|$ for these conservative problems. However, when $\gamma \neq 0$ the problem is generally non-conservative, and one may get responses which are non-intuitive from a physical point of view. This is shown in Fig. 8.10 for a "follower force" problem treated in (Pedersen 1977a) and other papers referred in this reference.

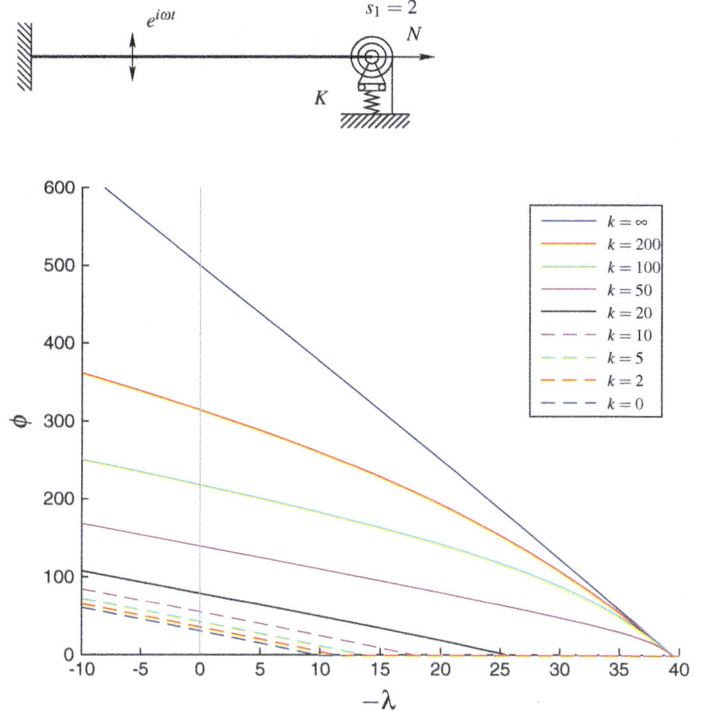

Fig. 8.9 Lowest eigenfrequency as a function of column force. The results are obtained by Newton–Raphson iterations for the case in Fig. 8.1h, here with $s_1 = 2$

Note in Fig. 8.10 that for $\gamma < 0.5$ a zero frequency is not obtained. The point of flutter instability ($\omega \neq 0$) is not shown as they are found for $|-\lambda| > 15$.

For a specific value of the follower parameter γ, the result in Fig. 8.11 is obtained by Newton–Raphson iterations for increasing column force λ to obtain the start of a complex solution for ϕ^2, i.e., start of an eigenvalue with $\alpha > 0$ and $\omega \neq 0$. Theory and more extended results similar to those in Fig. 8.11 may be seen in (Pedersen 1977a). This also includes the influence of linear external damping.

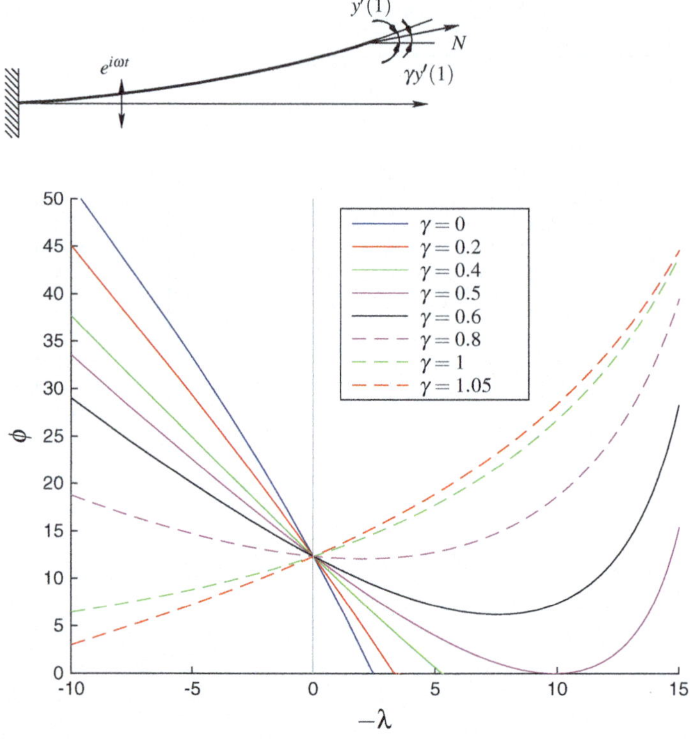

Fig. 8.10 Lowest eigenfrequency as a function of column force. The results are obtained by Newton–Raphson iterations for the last case in Fig. 8.11 but with factor $1 - \gamma$ originally in (8.5), i.e., the characteristic equation is $(1 - \gamma)\lambda D_5 + D_6 = 0$

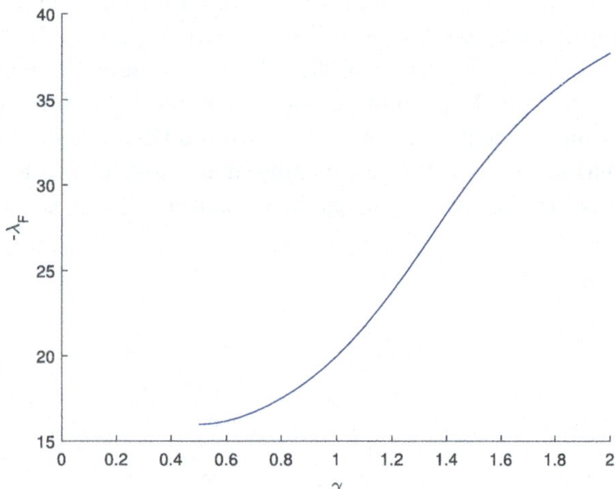

Fig. 8.11 Non-dimensional force at the free end λ_F that initiates flutter, as a function of the follower parameter γ

Chapter 9
Dynamic Stability Formulation

A cantilever loaded at its free end is used to introduce the subject of dynamic stability, including non-conservative axial external force such as the case presented at the end of Chap. 8. The essential difference between divergence instability with zero eigenfrequency and flutter instability, i.e., vibration with increasing amplitude is discussed. The limitation of formulations with a single degree of freedom is also presented.

9.1 One and Two Degrees of Freedom

We analyze simple one and two degrees of freedom systems with conservative as well as with non-conservative loads. Almost the same models are treated in (Panovko and Gubanova 1964) (pp. 57–74) with the history of the problem (pp. 72–74) and reference to the original papers by (Nikolai 1928), (Beck 1952), and (Dzhanelidze 1958). Here, we give a formulation in terms of flexibilities in order to give the presentation some generality besides the specific cantilever columns, which are discussed in detail.

References 1928–1977

Studies of a number of alternative models with few degrees of freedom are still reported. Specifically in relation to the behavior of the degenerated one degree of freedom models, we may mention the papers by (Lee and Reissner 1975) and (Neer and Baruch 1977). The aim is to show how much additional information a dynamic formulation gives in relation to that supplied by a static stability formulation. A secondary goal is to show some consequences of dealing only with one degree of freedom models.

© Springer International Publishing AG, part of Springer Nature 2018
S. L. Wiggers and P. Pedersen, *Structural Stability and Vibration*, Springer Tracts in Mechanical Engineering, https://doi.org/10.1007/978-3-319-72721-9_9

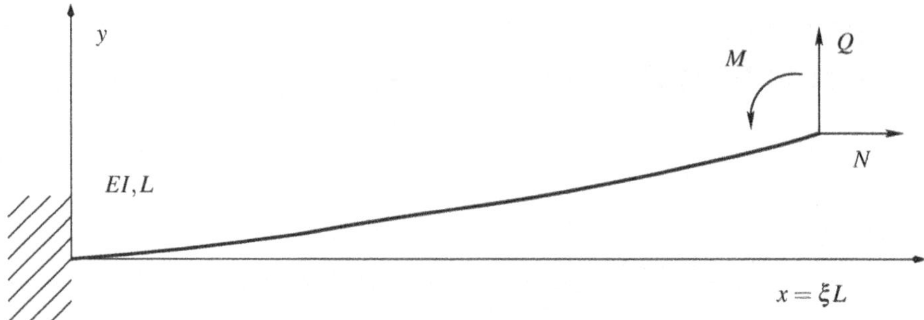

Fig. 9.1 Elementary case of a tip-loaded cantilever

9.2 An Elementary Beam-Column Case

Tip-loaded cantilever

A tip-loaded cantilever is shown in Fig. 9.1. This beam-column is slender and uniform with length L and bending stiffness EI. The beam loads are a transverse force Q in the y-direction and a couple M, anti-clockwise positive. The column load is an axial force N in the x-direction, i.e., positive in "tension."

Non-homogeneous problem

The well-known differential equation for this model is

$$y''''(\xi) - \lambda y''(\xi) \equiv 0 \ \text{ with definitions}$$

$$\xi := \frac{x}{L}, \qquad \lambda := \frac{NL^2}{EI}, \qquad y' := \frac{\partial y}{\partial \xi} \tag{9.1}$$

including the definition of the non-dimensional position $0 \le \xi \le 1$ and the non-dimensional force λ. The boundary conditions for the model are

$$y(0) = y'(0) = 0, \ \ y''(1) = \frac{ML^2}{EI}, \ \ y'''(1) - \lambda y'(1) = -\frac{QL^3}{EI} \tag{9.2}$$

Exact solutions to the mathematical problem (9.1)–(9.2) are given in many textbooks and will not be shown here because we are only interested in the relations between the tip ($\xi = 1$) displacements $y(1)$, $y'(1)$ and the tip loads Q, M, and N (N by λ).

Flexibilities

In matrix notation, the result is

$$\left\{ \begin{array}{c} y(1) \\ y'(1) \end{array} \right\} = [F] \left\{ \begin{array}{c} \frac{QL^3}{EI} \\ \frac{ML^2}{EI} \end{array} \right\} = \left[\begin{array}{cc} f_{11} & f_{12} \\ f_{21} & f_{22} \end{array} \right] \left\{ \begin{array}{c} \frac{QL^3}{EI} \\ \frac{ML^2}{EI} \end{array} \right\} \tag{9.3}$$

with

$$f_{11} := \frac{\sqrt{-\lambda}(\sin\sqrt{-\lambda} - \sqrt{-\lambda}\cos\sqrt{-\lambda})}{\lambda^2 \cos\sqrt{-\lambda}} = \frac{-\sqrt{\lambda}(\sinh\sqrt{\lambda} - \sqrt{\lambda}\cosh\sqrt{\lambda})}{\lambda^2 \cosh\sqrt{\lambda}}$$

$$f_{22} := \frac{-\lambda\sqrt{-\lambda}\sin\sqrt{-\lambda}}{\lambda^2 \cos\sqrt{-\lambda}} = \frac{\lambda\sqrt{\lambda}\sinh\sqrt{\lambda}}{\lambda^2 \cosh\sqrt{\lambda}}$$

$$f_{12} = f_{21} := \frac{-\lambda(1 - \cos\sqrt{-\lambda})}{\lambda^2 \cos\sqrt{-\lambda}} = \frac{-\lambda(1 - \cosh\sqrt{\lambda})}{\lambda^2 \cosh\sqrt{\lambda}}$$

$$f_D = f_{11}f_{22} - f_{12}^2 =$$

$$\frac{2(1 - \cos\sqrt{-\lambda}) - \sqrt{-\lambda}\sin\sqrt{-\lambda}}{\lambda^2 \cos\sqrt{-\lambda}} = \frac{2(1 - \cosh\sqrt{\lambda}) + \sqrt{\lambda}\sinh\sqrt{\lambda}}{\lambda^2 \cosh\sqrt{\lambda}} \tag{9.4}$$

where the important matrix determinant f_D is also presented.

Critical column force

Note that all flexibilities are real for all values of λ and that the diagonal flexibilities f_{11} and f_{22} may be negative for some values of λ. When $\cos\sqrt{-\lambda}$ approaches zero, these flexibilities approach $\pm\infty$, and the corresponding λ values are termed the critical loads, the buckling loads, or the Euler loads. The solutions to $\cos\sqrt{-\lambda} = 0$ are

$$\lambda_C = \frac{N_C L^2}{EI} = -\left(\frac{\pi}{2} + n\pi\right) \quad \text{for } n = 0, 1, 2, \ldots \tag{9.5}$$

Note that this gives no information about the behavior below and above these critical values. Furthermore, we shall see that in relation to the question of stability, only the value corresponding to $n = 1$ is of interest.

Figure 9.2 shows the flexibilities, and from these curves, we can directly determine the stability quantitatively as we shall see in the following sections. Before ending this section, we also list the stiffnesses, defined by

Fig. 9.2 Flexibilities and
flexibility determinant
defined by (9.4) as a function
of axial load

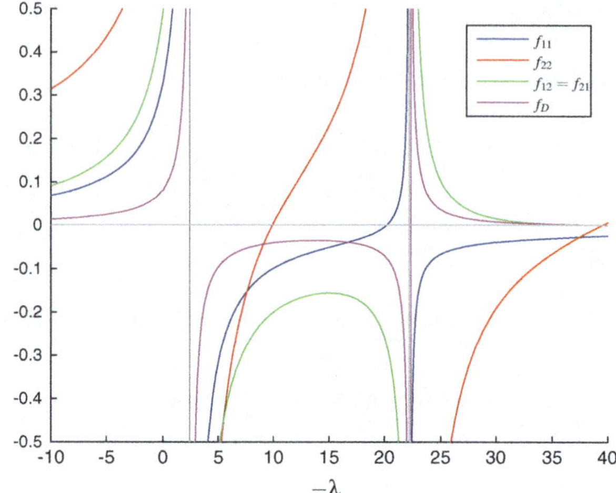

Fig. 9.3 Stiffnesses and
stiffness determinant defined
by (9.7) as a function of axial
load

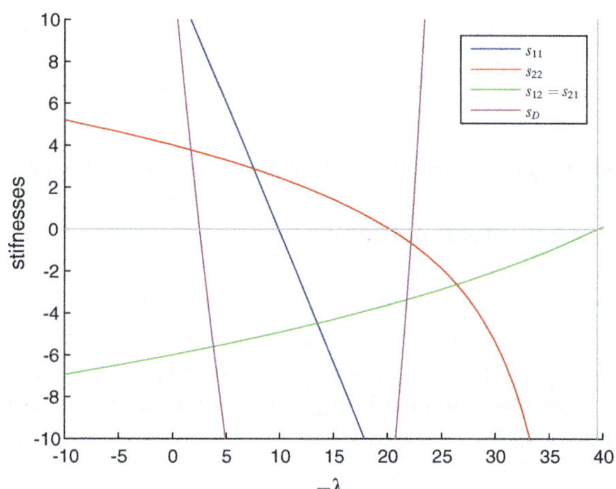

$$\left\{ \begin{array}{c} \frac{QL^3}{EI} \\ \frac{ML^2}{EI} \end{array} \right\} = [S] \left\{ \begin{array}{c} y(1) \\ y'(1) \end{array} \right\} = \left[\begin{array}{cc} s_{11} & s_{12} \\ s_{21} & s_{22} \end{array} \right] \left\{ \begin{array}{c} y(1) \\ y'(1) \end{array} \right\} \tag{9.6}$$

and thus obtained by inversion of $[F]$.

Stiffnesses

The real value expressions are:

$$
s_{11} := \frac{-\lambda\sqrt{-\lambda}\,\sin\sqrt{-\lambda}}{2(1-\cos\sqrt{-\lambda})-\sqrt{-\lambda}\,\sin\sqrt{-\lambda}} = \frac{\lambda\sqrt{\lambda}\,\sinh\sqrt{\lambda}}{2(1-\cosh\sqrt{\lambda})+\sqrt{\lambda}\,\sinh\sqrt{\lambda}}
$$

$$
s_{22} := \frac{\sqrt{-\lambda}(\sin\sqrt{-\lambda}-\sqrt{-\lambda}\cos\sqrt{-\lambda})}{2(1-\cos\sqrt{-\lambda})-\sqrt{-\lambda}\,\sin\sqrt{-\lambda}} = \frac{-\sqrt{\lambda}(\sinh\sqrt{\lambda}-\sqrt{\lambda}\cosh\sqrt{\lambda})}{2(1-\cosh\sqrt{\lambda})+\sqrt{\lambda}\,\sinh\sqrt{\lambda}}
$$

$$
s_{12} := \frac{\lambda(1-\cos\sqrt{-\lambda})}{2(1-\cos\sqrt{-\lambda})-\sqrt{-\lambda}\,\sin\sqrt{-\lambda}} = \frac{\lambda(1-\cosh\sqrt{\lambda})}{2(1-\cosh\sqrt{\lambda})+\sqrt{\lambda}\,\sinh\sqrt{\lambda}}
$$

$$
s_D := s_{11}s_{22} - s_{12}^2 = f_D^{-1} \tag{9.7}
$$

and Fig. 9.3 shows the dependence on λ.

9.3 Column with a Point Mass

The cantilever column shown in Fig. 9.4 is loaded with a "dead" load P, i.e., a conservative load which is not influenced by the displacements of its point of application. The buckling load P_C for this column is

$$
P_C = \frac{\pi^2}{4}\frac{EI}{L^2} \tag{9.8}
$$

in agreement with (9.5) for the case of $n = 1$. This classic result is obtained by pure statics and is thus not influenced by the mass distribution. Therefore, we get the same buckling load with a point mass m at the tip, as also shown in Fig. 9.4.

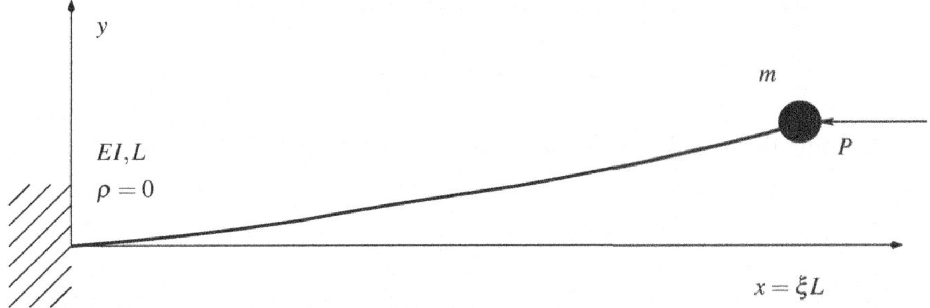

Fig. 9.4 Column with a "dead" load P and a point mass m at the tip

Problem of dynamic stability

Let us now assume this point mass to be the only important mass of the column and then formulate the problem of dynamic stability, i.e.,

$$\text{"THE COLUMN IS VIBRATING—ARE THE VIBRATIONS STABLE ?"} \quad (9.9)$$

Based on the d'Alembert principle, we convert the dynamic problem into a static one by adding the force Q defined by

$$Q = -m\frac{\partial^2 y}{\partial t^2} \quad \text{evaluated at} \ \ \xi = 1 \tag{9.10}$$

with the direction of Q as shown in Fig. 9.1.

No complex frequency

In the exponential time function:

$$e^{(\alpha+i\omega)t} = e^{\alpha t}e^{i\omega t} = e^{\alpha t}(\cos \omega t + i \sin \omega t) \tag{9.11}$$

we prefer the complex argument in order to avoid the frequently used, but confusing term "complex frequency." The second factor $e^{i\omega t}$ describes a periodic (harmonic) time function with

$$\omega = \text{FREQUENCY PARAMETER} \tag{9.12}$$

and the first factor $e^{\alpha t}$ describes the stability with unlimited growth for $\alpha > 0$, disappearance for $\alpha < 0$, and critical value for $\alpha = 0$. Quantitative information about stability/instability is therefore given directly by

$$\alpha = \text{STABILITY PARAMETER} \tag{9.13}$$

Now, returning to our specific problem, from (9.10) and (9.11) we get

$$Q = -m(\alpha + i\omega)^2 y(1) \tag{9.14}$$

or, by defining the non-dimensional complex parameter μ^2

$$\frac{QL^3}{EI} = -\mu^2 y(1) \quad \text{with} \ \ \mu^2 := \frac{mL^3(\alpha + i\omega)^2}{EI} \tag{9.15}$$

From the elementary case result (9.3), we then with (9.15) and $M = 0$ obtain the result:

$$y(1) = f_{11} \frac{QL^3}{EI} = -f_{11} \mu^2 y(1) \tag{9.16}$$

which gives $\mu^2 = -(f_{11})^{-1}$ or, in terms of (stability, frequency)

$$(\alpha + i\omega)^2 = -\frac{EI}{mL^3} \frac{1}{f_{11}} \tag{9.17}$$

with the dependence on P implicit by f_{11} as listed in (9.4).

Solution by sign of f_{11}

Since f_{11} is a real quantity, we can only get the solution ($\alpha = 0, \omega \neq 0$) or ($\alpha \neq 0, \omega = 0$). The first corresponds to $f_{11} > 0$, which we have then proved to give HARMONIC VIBRATIONS with frequency

$$\omega = \pm\sqrt{\frac{EI}{mL^3 f_{11}}} \quad \text{for} \quad f_{11} > 0 \tag{9.18}$$

The second solution we get for $f_{11} < 0$, i.e.,

$$\alpha = \pm\sqrt{-\frac{EI}{mL^3 f_{11}}} \quad \text{for} \quad f_{11} < 0 \tag{9.19}$$

and as one of these solutions has positive α value, the column is unstable. This instability ($\alpha > 0, \omega = 0$) is a static instability, termed DIVERGENCE.

The other kind of instability ($\alpha > 0, \omega \neq 0$) is termed FLUTTER, but as we have just seen, the model in Fig. 9.4 cannot exhibit flutter. (We need more than one degree of freedom to describe the flutter instability.)

The answer to our primary question (9.9) can now be read directly from Fig. 9.2, which confirms the buckling load P_C from (9.8) because

$$f_{11} > 0 \text{ for } P < P_C \qquad f_{11} < 0 \text{ for } P > P_C \tag{9.20}$$

Furthermore, $f_{11} = \pm\infty$ for $P = P_C$ shows that the buckling load corresponds to zero eigenfrequency.

Note that the dynamic analysis gives information about the behavior in load domains: For $P < P_C$, we have harmonic vibrations with the frequency ω determined by (9.18), and for $P > P_C$, we have divergent instability with the positive exponential factor α determined by (9.19).

Information on load domains

A non-physical result follows from this very degenerated model, which has only a point mass to model inertia. Referring to Fig. 9.2, we note that f_{11} is again positive in a narrow interval before $-\lambda \simeq 22$. In these post-buckling domains, we are in conflict with our basic assumptions. We shall therefore leave the present too simple model and analyze a two degrees of freedom model.

9.4 An Improved Dynamic Column Model

Figure 9.5 shows an improved model for analysis of dynamic stability. The improvement in relation to the model in Fig. 9.4 is that we take rotational inertia into account. The mass moment of inertia J is

$$J = mL^2\zeta^2 \tag{9.21}$$

by which the non-dimensional factor ζ is defined (if not directly given as in Fig. 9.5.

The d'Alembert principle then gives two beam end loads at the tip: a transverse force Q and a moment M by

$$Q = -m\frac{\partial^2 y}{\partial t^2}(1) = -m(\alpha + i\omega)^2 y(1)$$

$$M = -mL^2\zeta^2\frac{\partial^2 y'}{\partial t^2}\frac{1}{L} = -mL\zeta^2(\alpha + i\omega)^2 y'(1) \tag{9.22}$$

Inserting this in the elementary case (9.3), we get

$$\left\{\begin{array}{c} y(1) \\ y'(1) \end{array}\right\} = -[F]\mu^2\left\{\begin{array}{c} y(1) \\ \zeta^2 y'(1) \end{array}\right\} \quad \text{with} \quad \mu^2 := \frac{mL^3(\alpha + i\omega)^2}{EI} \tag{9.23}$$

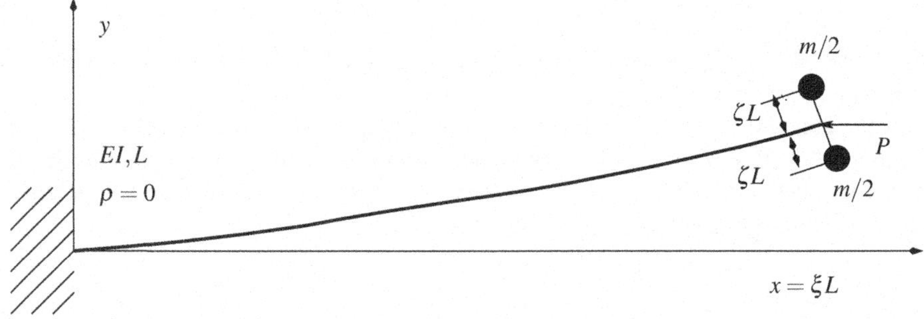

Fig. 9.5 Column model with rotational inertia

which we write in homogeneous form

$$
\begin{bmatrix} 1 + \mu^2 f_{11} & \zeta^2 \mu^2 f_{12} \\ \mu^2 f_{12} & 1 + \zeta^2 \mu^2 f_{22} \end{bmatrix} \begin{Bmatrix} y(1) \\ y'(1) \end{Bmatrix} = \begin{Bmatrix} 0 \\ 0 \end{Bmatrix}
\tag{9.24}
$$

Homogeneous formulation

A nonzero solution is obtained when the determinant of the coefficient matrix is zero, i.e., when

$$
\zeta^2 (f_{11} f_{22} - f_{12}^2) \mu^4 + (f_{11} + \zeta^2 f_{22}) \mu^2 + 1 = 0
\tag{9.25}
$$

with the solution (9.17) for $\zeta = 0$. For $\zeta \neq 0$, too, the solution gives a real μ^2 because the discriminant is non-negative

$$
4\zeta^2 f_{12}^2 + (f_{11} + \zeta^2 f_{22})^2 \geq 0
\tag{9.26}
$$

No flutter possible

From this, we directly conclude that the model cannot exhibit flutter, i.e., not ($\alpha >$ 0, $\omega \neq 0$).

The sign of μ^2 then returns harmonic vibrations for $\mu^2 < 0$ and divergent instability for $\mu^2 > 0$, which by solving (9.25) or directly from the criteria in Chap. 15 gives

$$
\text{harmonic vibrations for } \quad f_{11} f_{22} - f_{12}^2 = f_D > 0 \ \text{ and } \ f_{11} + \zeta^2 f_{22} > 0
$$
$$
\text{divergent instability for } \quad f_D < 0 \ \text{ or } \ f_{11} + \zeta^2 f_{22} < 0
\tag{9.27}
$$

With the graphical display in Fig. 9.2, this then shows complete agreement with the classic result that the first buckling load (9.8) is the load which separates harmonic vibrations from divergent instability. The non-physical results for the too simple model with only one degree of freedom are no longer found.

Complete information—Limiting case

For a given ratio of mass moment to mass, i.e., ζ^2, we can solve (9.25) to obtain $\mu^2 = \mu^2(\lambda)$, but we shall only give the asymptotic results for $\mu^2 \to \infty$ for comparison with result (9.17). With the μ^2 definition (9.15), we get

$$
(\alpha + i\omega)^2 = -\frac{EI}{mL^3} \frac{f_{22}}{f_D} \quad \text{for } \mu^2 \to \infty
\tag{9.28}
$$

9.5 Non-conservative Column Load

We now return to the dynamic model with only one degree of freedom. However, we release the assumption of a "dead" column load and treat the extended column load shown in Fig. 9.6. This column load is termed a "partial follower force" with its direction depending on the column end rotation by $y'(1)$ and the "follower parameter" γ. The problem solved in Sect. 9.3 corresponds to $\gamma = 0$ and the problem of a tangential load corresponds to $\gamma = 1$.

By the assumption of small displacements, we find the (x, y)-components of the force N to be $(N, N\gamma \frac{y'(1)}{L})$. Thus, the total transverse force Q will be $Q = -m\ddot{y}(1) + N\gamma \frac{y'(1)}{L}$ or

$$\frac{QL^3}{EI} = -\mu^2 y(1) + \lambda\gamma y'(1) \ \text{ with the definitions}$$

$$\mu^2 := \frac{mL^3(\alpha + i\omega)^2}{EI} \ \text{ and } \ \lambda := \frac{NL^2}{EI} \tag{9.29}$$

Inserting this in the elementary case result (9.3) with $M = 0$, we have:

$$\left\{ \begin{array}{c} y(1) \\ y'(1) \end{array} \right\} = [F] \left\{ \begin{array}{c} -\mu^2 y(1) + \lambda\gamma y'(1) \\ 0 \end{array} \right\} \tag{9.30}$$

which we write in homogeneous form

$$\begin{bmatrix} 1 + \mu^2 f_{11} & -\gamma\lambda f_{11} \\ \mu^2 f_{12} & 1 - \gamma\lambda f_{12} \end{bmatrix} \left\{ \begin{array}{c} y(1) \\ y'(1) \end{array} \right\} = \left\{ \begin{array}{c} 0 \\ 0 \end{array} \right\} \tag{9.31}$$

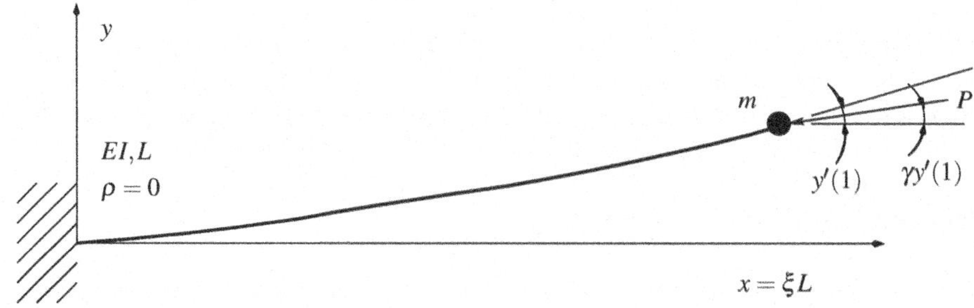

Fig. 9.6 Column model with partial follower force P and a point mass m at the tip

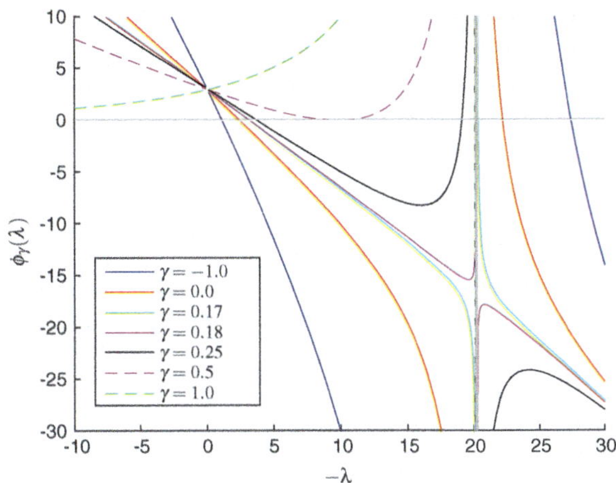

Fig. 9.7 Function ϕ_γ as a function of λ defined in (9.33). Positive values imply harmonic vibrations, and negative values imply divergent instability. The level of non-conservative force γ is treated as parameter

The determining condition for a nonzero solution then gives

$$(\alpha + i\omega)^2 = -\frac{EI}{mL^3}\frac{1 - \gamma\lambda f_{12}}{f_{11}} \tag{9.32}$$

which, for $\gamma = 0$ (dead load), agrees with our earlier result (9.17). Primarily, we note that the right-hand side is always a real quantity and that our model thus cannot exhibit flutter instability.

Sign change nominator or denominator

The question of harmonic vibration or divergent instability is then answered directly by the sign of the important function $1 - \gamma\lambda f_{12}$, and this sign can be changed by the nominator or by the denominator. Introducing our specific flexibilities (9.4), we get the well-known result

$$(\alpha + i\omega)^2 = -\frac{EI}{mL^3}\frac{\lambda^2((1-\gamma)\cos\sqrt{-\lambda} + \gamma)}{\sqrt{-\lambda}(\sin\sqrt{-\lambda} - \sqrt{-\lambda}\cos\sqrt{-\lambda})} =: -\frac{EI}{mL^3}\phi_\gamma(\lambda) \tag{9.33}$$

with its less well-known graphical display, shown in Fig. 9.7.

The parameter γ controls the non-conservative force component relative to the Euler component, and we see how strongly the behavior (α, ω) depends on γ. The common singularity at $-\lambda = \simeq 20.2$ (found for $\tan\sqrt{-\lambda} = \sqrt{-\lambda}$) is only described by the dynamic formulation. Now, the physical interpretation of Fig. 9.7 is as follows:

Weakly non-conservative

For $\gamma \leq 0.175$, which we may describe as weakly non-conservative, the result is harmonic vibrations for axial forces limited by specific flexibilities (9.4), and we get the well-known result

$$(1 - \gamma)\cos\sqrt{-\lambda} + \gamma > 0 \tag{9.34}$$

and increasing the compressive axial force above this, we get divergent instability. The narrow positive interval after the singularity we interpret as being non-physical in the same sense as proved for the conservative load in Sects. 9.3 and 9.4.

Moderate non-conservative

In the range of moderate non-conservative loads, $0.175 < \gamma < 0.5$, the result is harmonic vibrations until the axial load determined by: $(1 - \gamma)\cos\sqrt{-\lambda} + \gamma = 0$, and then, a divergent instability, which is, however, stabilized again before the singularity is reached. The stabilization load is determined by the second solution to $(1 - \gamma)\cos\sqrt{-\lambda} + \gamma = 0$. The load domain in which the divergent instability is active narrows down when γ increases and degenerates to a point for $\gamma = 0$. Thus, the critical load from $(1 - \gamma)\cos\sqrt{-\lambda} + \gamma = 0$ is very special in the sense that we get harmonic vibrations below as well as above this value. Not until the singularity $\lambda \simeq 20.2$ do we get a clear instability. Note also that for $0.175 < \gamma < 0.5$, the "degree of instability" (value of α) is limited.

Strongly non-conservative

Finally, with loads that are strongly non-conservative, $\gamma > 0.5$, we see the limitations of a static stability analysis. The nominator of (9.33), $(1 - \gamma)\cos\sqrt{-\lambda} + \gamma$, will never be zero. This does not mean that the model will always be stable, but that the formulation is too restricted to locate the instability. It may seem paradoxical that a divergent instability at $-\lambda \simeq 20.2$ cannot be located by a static formulation. We may explain this by saying that it takes an imperfection to locate the instability and that the inertia force serves this purpose.

As seen from Fig. 9.7, the divergent instability for $\gamma > 0.5$ is independent of the follower parameter γ and the divergent instability after $-\lambda \simeq 20.2$ is reached through infinite frequency ($\omega = \infty$) and not through vanishing frequency ($\omega = 0$).

Chapter 10
Stability of 2D Frames

Five cases of stability analysis for 2D frames are presented in detail.

Then, a solution procedure for these rather simple frames is presented as a kind of "cookbook." A case (Roorda–Koiter frame) of critical imperfection analysis is shown with solution obtained by applying Berry functions. The necessary statics to formulate this problem is a good exercise.

10.1 Frames

Figure 10.1 shows some simple 2D frames, each of which are a combination of beam-columns. The in-plane buckling force for these frames may be determined by applying the results in Chap. 7, some rather directly and others with a procedure (a cookbook) to be presented in this chapter. Finally in Sect. 10.5, a post-critical analysis is shown.

10.2 Only One Beam-Column in the Frame

The rather direct solution is for frames where only one of the uniform beams are subjected also to a column force. The remaining beams then act as end springs. Figure 10.1a shows the first example. The non-dimensional spring stiffness s_1 from beam ABC follows from results for simple beam theory (4.14) with $\lambda = 0$, $M_0 = M/2$, $\eta = 1$ and stiffness $S_1 = M/\theta_0$ giving

$$S_1 = \frac{s_1 EI}{L} = 2\frac{3Ei}{l} \quad \rightarrow \quad s_1 = \frac{6iL}{Il} \tag{10.1}$$

© Springer International Publishing AG, part of Springer Nature 2018
S. L. Wiggers and P. Pedersen, *Structural Stability and Vibration*, Springer Tracts in Mechanical Engineering, https://doi.org/10.1007/978-3-319-72721-9_10

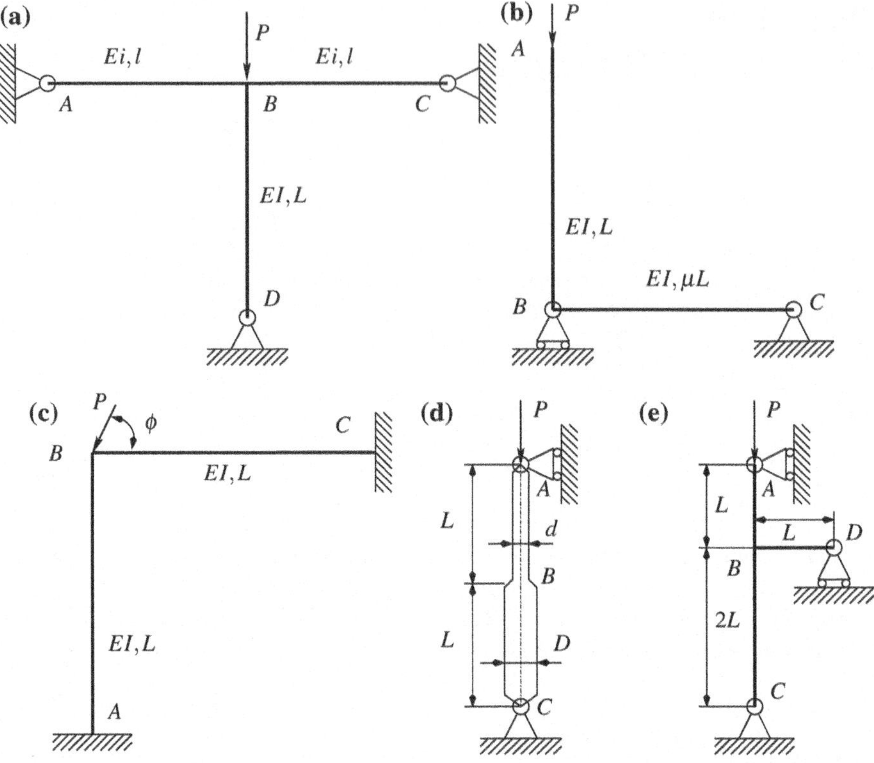

Fig. 10.1 Different simple 2D frame models to be analyzed in the present chapter

as the rotational stiffness for each of the horizontal beams is found from $\theta_1 = \frac{l}{Ei} M_1 \frac{1}{3}$ giving $\frac{M_1}{\theta_1} = \frac{3Ei}{l}$.

Result example a

The resulting transcendental equation from (4.20) with $s_0 = 0$ implies

$$\frac{1}{6} + \frac{iL}{Il} B_1 = 0 \tag{10.2}$$

that with numerical solution for $\frac{Il}{iL} = 4$ is extrapolated from Figs. 4.2 and 4.4 to

$$\frac{1}{6} + \frac{B_1}{4} = 0 \;\Rightarrow\; B_1 = -\frac{2}{3} \;\Rightarrow\; \lambda_C = \frac{N_C L^2}{EI} \simeq -12 \tag{10.3}$$

Result example b

Figure 10.1b shows a second frame example. The end spring stiffness s_0 from (4.14) with $\lambda = 0$, $M_0 = M$, $\eta = 1$, $S_0 = \frac{M}{\theta_0}$ gives

$$S_0 = \frac{s_0 EI}{L} = \frac{EI}{\mu L}\frac{1}{2} \quad \rightarrow \quad s_0 = \frac{1}{2\mu} \tag{10.4}$$

Further from (7.8) follows

$$\lambda B_1 + 1 + \frac{\mu \lambda}{3} = 0 \tag{10.5}$$

that for given numerical value of μ gives the non-dimensional buckling force λ_C. Limit solutions are

$$\lambda_C = -\pi^2/4 \ \text{ for } \ \mu = 0 \ \text{ and } \ \lambda_C \ \rightarrow \ 0 \ \text{ for } \ \mu \ \rightarrow \ \infty \tag{10.6}$$

10.3 Several Beam-Columns in the Frame

For frames combined of beams with more than one beam-column, the stiffness of these beam-columns is not constant which complicates the necessary calculations. Figure 10.1, Example c, is analyzed to illustrate this. Dividing the frame into two beam-columns AB and BC, the column forces just before buckling are $N_{AB} = -P \sin \phi$ and $N_{BC} = P \cos \phi$. With a rotation ψ of point B for static equilibrium, as shown in Fig. 10.2, the elementary case (4.14) gives

Result example c

$$\text{For beam-column } \ AB := \lambda_{AB} = \frac{-PL^2 \sin \phi}{EI}$$

$$\theta_{AB} = 0 = \frac{L}{EI}(M_A B_1(\lambda_{AB}) + M_B B_0(\lambda_{AB})) \quad \rightarrow \quad M_A = \frac{-M_B B_0(\lambda_{AB})}{B_1(\lambda_{AB})}$$

$$\theta_{BA} = \psi = \frac{L}{EI} M_B \left(\frac{-B_0^2(\lambda_{AB})}{B_1(\lambda_{AB})} + B_1(\lambda_{AB}) \right)$$

$$\text{For beam-column } \ BC := \lambda_{BC} = \frac{PL^2 \cos \phi}{EI}$$

$$\theta_{CB} = 0 \quad \rightarrow \quad M_C = \frac{M_B B_0(\lambda_{BC})}{B_1(\lambda_{BC})}$$

$$\theta_{BC} = \psi = \frac{-L}{EI} M_B \left(\frac{-B_0^2(\lambda_{BC})}{B_1(\lambda_{BC})} + B_1(\lambda_{BC}) \right) \tag{10.7}$$

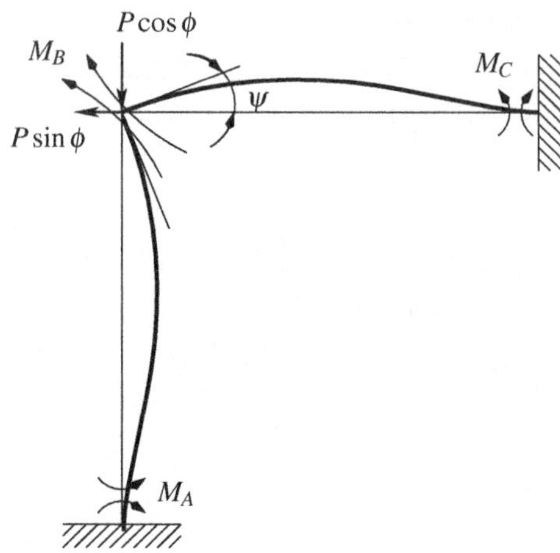

Fig. 10.2 Example c displaced for static equilibrium

Eliminating ψ a transcendental equation is obtained

$$\frac{B_1^2(\lambda_{AB}) - B_0^2(\lambda_{AB})}{B_1(\lambda_{AB})} = -\frac{B_1^2(\lambda_{BC}) - B_0^2(\lambda_{BC})}{B_1(\lambda_{BC})}$$
$$\text{with}\quad \lambda_{BC} = -\lambda_{AB}\cot\phi \tag{10.8}$$

and this may be solved numerically by the Newton–Raphson method to obtain $P_C = P_C(\phi)$.

Result example d

The beam-column with change of diameter in Fig. 10.1d is modeled as two beam-columns with different non-dimensional column forces λ_{AB} and λ_{BC}.

$$\lambda = \lambda_{AB} = \frac{-PL^2}{EI(d)} \quad\text{and}\quad \lambda_{BC} = \frac{-PL^2}{EI(D)} = \lambda\left(\frac{d}{D}\right)^4 \tag{10.9}$$

From Fig. 10.3 is seen the chosen displacement parameters δ and ψ, and the angles $\theta_{BA} = \delta/L + \psi$ and $-\theta_{BC} = \delta/L - \psi$. Elementary case (4.14) for the two beam-columns with $M_A = M_C = 0$ gives

$$\theta_{BA} = \frac{L}{EI}M_B B_1(\lambda), \quad \theta_{BC} = \frac{L}{EI}\left(\frac{d}{D}\right)^4(-M_B)B_1\left(\lambda\left(\frac{d}{D}\right)^4\right) \tag{10.10}$$

and with $M_B = P\delta$ from statics the homogeneous system of equations are

$$\frac{\delta}{L} + \psi = \frac{L}{EI}P\delta B_1(\lambda) = -\lambda\frac{\delta}{L}B_1(\lambda)$$
$$\frac{-\delta}{L} + \psi = \lambda\frac{\delta}{L}\left(\frac{d}{D}\right)^4 B_1\left(\lambda\left(\frac{d}{D}\right)^4\right) \tag{10.11}$$

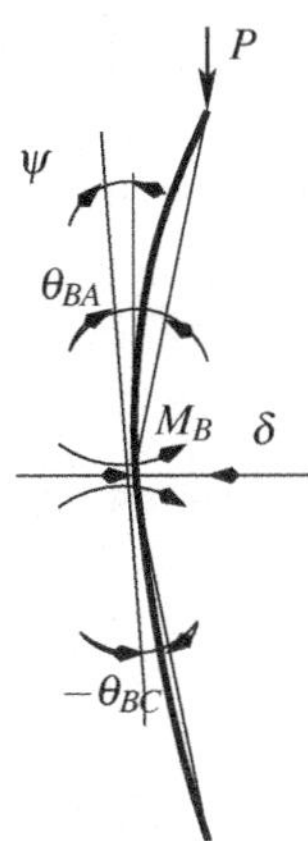

Fig. 10.3 Example d displaced for static equilibrium

that only has a solution for $(\frac{\delta}{L}, \psi) \neq (0, 0)$ if

$$\lambda \left(B_1(\lambda) + (\frac{d}{D})^4 B_1(\lambda(\frac{d}{D})^4) \right) = -2 \tag{10.12}$$

with known solution for $d = D$, i.e., $\lambda = -\pi^2/4$ (note total length is $2L$).

10.4 Solution Procedure ("cookbook")

Based on the examples in Figs. 10.1c and d, a "cookbook" is set up for stability analysis of such frames. It will be applied to the fifth example.

General solution procedure

(a) The frame is divided into beam-columns for which, elementary solutions are available.
(b) The column force in each beam-column is determined. These non-dimensional forces λ can be determined in the non-displaced structure/system.
(c) Displacement parameters are chosen, i.e., end translations and/or end rotations.
(d) Geometrical relations between chosen displacement parameters and the parameters of the elementary case, say θ_0, θ_1.
(e) Statical relations between loads—chosen displacement parameters and moments in the elementary case, say M_0, M_1.
(f) Physical relations from elementary cases.
(g) Combination to a homogeneous system of equations with chosen displacements parameters as unknown.
(h) The determinant for the coefficient matrix must be zero for determination of the buckling force. This will normally result in a transcendental equation.

(i) Numerical solution of this transcendental equation, say by the Newton–Raphson method.

(j) Test for dimensionality and being reasonable.

The problem in Fig. 10.1e is more complicated than the first four examples. A solution following the listed procedure is shown. Figure 10.4 shows this fifth example in displaced mode.

Result example e—A more complicated example

(a) The beam-columns are AB, BC, BD.

(b) Column forces in non-displaced frame are $N_{AB} = -P$, $N_{BC} = -P$, and $N_{BD} = 0$.

(c) The chosen displacement parameters are translation δ and rotation ψ as defined in Fig. 10.4.

(d) The end angles θ_{BA}, θ_{BC}, and θ_{BD} are defined from chord to the end tangent in the displaced frame, positive anti clockwise. The relations to δ, ψ are $\theta_{BA} = \psi - \frac{\delta}{L}$, $\theta_{BA} = \psi + \frac{\delta}{2L}$, and $\theta_{BD} = \psi$.

(e) Equilibrium for internal moments at point B gives $M_{BA} + M_{BC} + M_{BD} = 0$. For the individual beam-columns: $M_{BA} = RL - P\delta$, $M_{BC} = 2RL + (P - H)\delta$, and $M_{BD} = -HL$. By eliminations of H and R the result is $2M_{BA} - M_{BC} + M_{BD}\frac{\delta}{L} = -3P\delta$.

(f) Elementary case (4.8) gives $\theta_{BA} = \frac{L}{EI}B_1(\lambda)M_{BA}$, $\theta_{BC} = \frac{2L}{EI}B_1(4\lambda)M_{BC}$, and $\theta_{BD} = \frac{L}{EI}\frac{M_{BD}}{3}$.

(g) Inserting the results from (d) in (f).

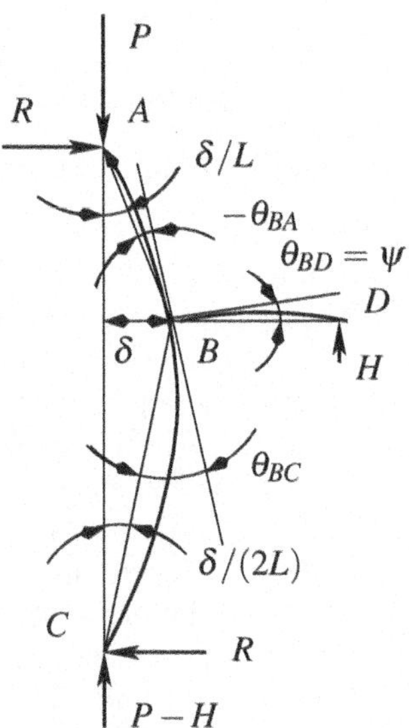

Fig. 10.4 Example e displaced for static equilibrium

(h) Neglecting the squared order term $\frac{\psi}{2}\frac{\delta}{L}$, the determinant condition is $D = 12\lambda B_1(\lambda)B_1(4\lambda) + (1 + 2\lambda)B_1(\lambda) + (\psi + 4\lambda)B_1(\lambda) + 3 = 0$.

(i) Numerical solution of this transcendental equation $D = 0$ by Newton–Raphson iterations to the final result $-\lambda_C = \frac{P_C L^2}{EI} \simeq 1.4$.

(j) Without beam BD, the buckling force is known to be $-\bar{\lambda}_C = \frac{P_C(3L)^2}{EI} = \pi^2 \rightarrow \frac{P_C L^2}{EI} = 1.1$ a value less than 1.4 as expected.

10.5 Post-Critical Imperfection Analysis

A frame (or in general a structure) may have a stable post-critical ($P > P_C$) behavior. For such structures, the critical column force does not imply immediate structural collapse.

Imperfection analysis

However, structures exist that may collapse before a critical column force is reached. These structures are named imperfection sensitive, and a collapse is depending on the actual imperfections (art and size)

Literature on the subject is extensive, and the present section only analyze a specific 2D frame example, named the Roorda–Koiter frame in agreement with (Roorda 1965), (Koiter 1967), and (Roorda and Chilver 1970). Figure 10.5 shows this frame, non-displaced as well as displaced for the necessary static analysis.

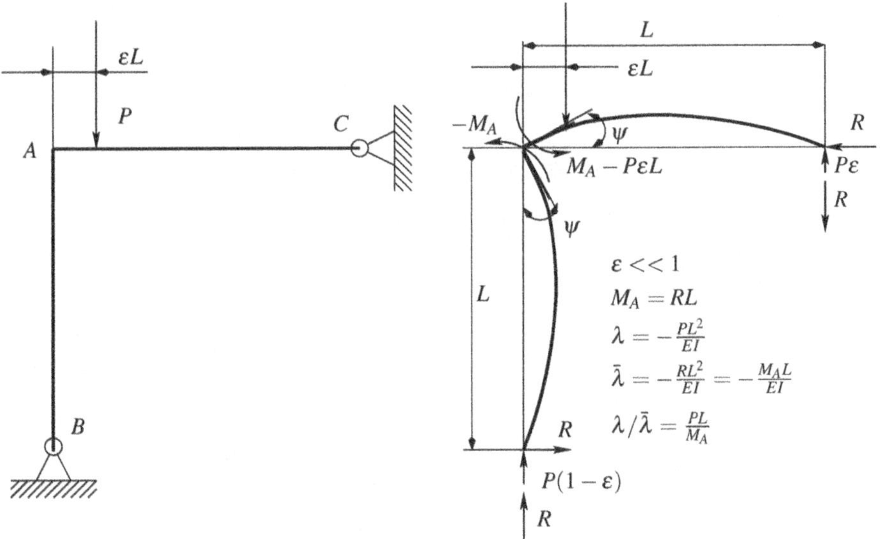

Fig. 10.5 The Roorda–Koiter frame to exemplify post-critical imperfection analysis. Note that the boundary moment at A for beam-column AC is $M_A - P\varepsilon L$

For beam-column AB with $M_B = 0$, the elementary case (4.8) gives

$$\psi = \frac{L}{EI}(-M_A)B_1((1-\varepsilon)\lambda + \bar{\lambda})$$

$$\text{with} \quad \lambda = \frac{-PL^2}{EI}, \quad \bar{\lambda} = \frac{-RL^2}{EI} = \frac{-M_A L}{EI} \tag{10.13}$$

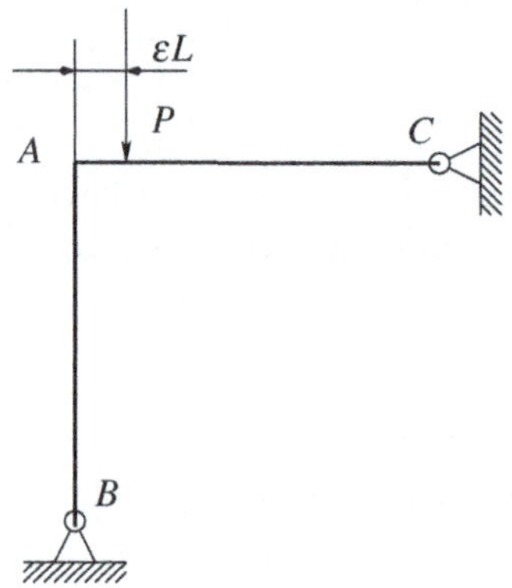

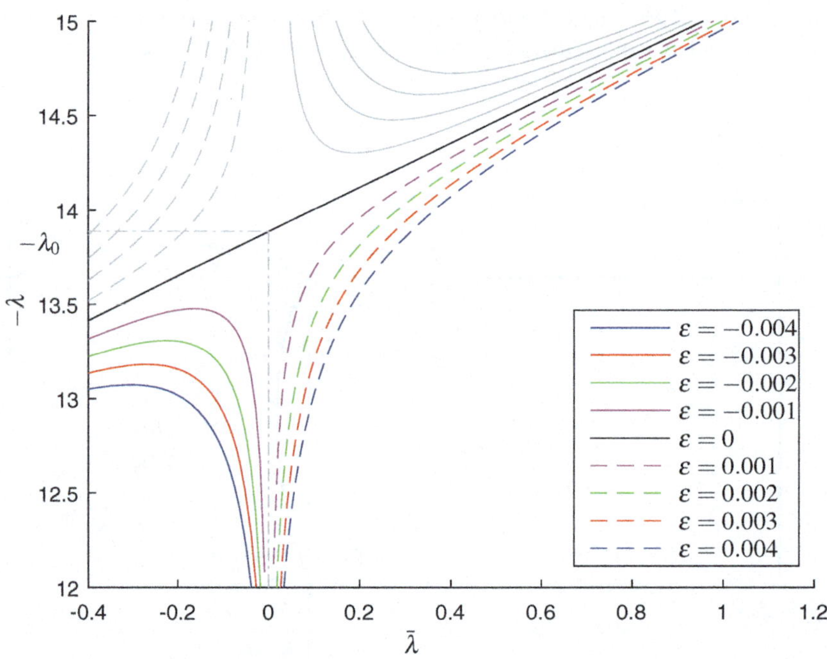

Fig. 10.6 Column force λ in AB as a function of column force $\bar{\lambda}$ in AB. For $\bar{\lambda} = 0$, the critical non-dimensional column force is $\lambda_0 \simeq -13.89$

and $\varepsilon << 1$ is assumed. The other beam-column AC with $M_C = 0$ gives

$$\psi = \frac{L}{EI}(M_A - P\varepsilon L)B_1(\bar{\lambda}) = \frac{L}{EI}M_A(1 - \varepsilon\frac{\lambda}{\bar{\lambda}})B_1(\bar{\lambda}) \qquad (10.14)$$

In displaced mode $M_A \neq 0$ with (10.13) minus (10.14) giving a transcendental equation to determine λ as a function of $\bar{\lambda}$ with ε as parameter.

$$B_1\left((1-\varepsilon)\lambda + \bar{\lambda}\right) + (1 - \varepsilon\frac{\lambda}{\bar{\lambda}})B_1(\bar{\lambda}) = 0 \qquad (10.15)$$

The displaced angle ψ is determined as the sum of (10.13) and (10.14)

$$2\psi = \bar{\lambda}B_1\left((1-\varepsilon)\lambda + \bar{\lambda}\right) - \bar{\lambda}(1 - \varepsilon\frac{\lambda}{\bar{\lambda}})B_1(\bar{\lambda}) \qquad (10.16)$$

or with (10.15)

$$\psi = (\varepsilon\lambda - \bar{\lambda})B_1(\bar{\lambda}) = \bar{\lambda}B_1\left((1-\varepsilon)\lambda + \bar{\lambda}\right) \qquad (10.17)$$

A numerical solution by Newton–Raphson method of (10.15) is the basis for Fig. 10.6 that shows $\lambda = \lambda(\bar{\lambda})$ with ε as parameter.

The final evaluated solution for $\psi = \psi(\frac{\lambda}{\lambda_0})$ is shown in Fig. 10.7 with ε as parameter, and λ_0 is non-dimensional critical force for $\varepsilon = 0$, being $\lambda_0 \simeq -13.89$.

Figure 10.7 **comments**

The comment to Fig. 10.7 are as follows:

- Only curves under the straight line corresponding to $\varepsilon = 0$ are of physical interest. These additional mathematical solutions are only shown with gray color.

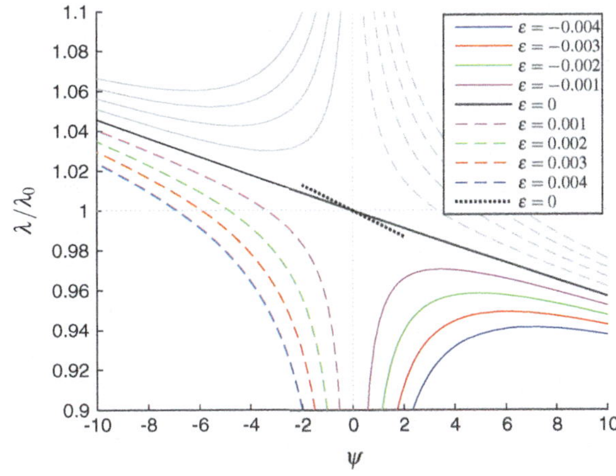

Fig. 10.7 Numerical results for simplified post-critical analysis of the Roorda–Koiter frame. Rotation ψ of point A as a function of relative force P, i.e., $\psi = \psi(\frac{\lambda}{\lambda_0})$ with ε as parameter. The critical non-dimensional force λ_0 for $\varepsilon = 0$ is $\lambda_0 \simeq -13.89$. Note, that the scale for ψ is in degrees

- Note, that the buckling behavior is not symmetric.
- For positive $\varepsilon > 0$, compressive forces larger than the critical compressive column force corresponding to $\varepsilon = 0$ are possible. In Fig. 10.6, this is seen to correspond to tensile force in column AC.
- For negative $\varepsilon < 0$, the collapse force is smaller than the critical compressive column force and increasingly smaller for numerically larger imperfection.
- For $\varepsilon = 0$, a dotted line is included corresponding to the results in the references mentioned initial in the present section. The difference relative to the shown result is due to the fact that the present simple analysis does not account for translational displacements of point A.

Chapter 11
Buckling Stresses, Material Nonlinearity, and Beam Modeling

Different concepts are introduced to increase the use of earlier presented results to other cross sections and other BC. The concepts of inertia radius, equivalent length, and slenderness ratio are important for beam-columns. Longitudinal stresses at buckling initiation are determined, and the approach for maximum stress in a beam-column is described. Influence from material nonlinearity and relation to design approaches are introduced. Finally, the improved Timoshenko beam theory and its influence on column buckling are described.

11.1 Various Concepts

Concept of inertia radius

In order not only to study a specific cross section and specific BC, the concept of slenderness ratio Γ and the concept of equivalent length l_e are introduced. The relation between cross-sectional moment of inertia I and area A is assumed to be given by the cross-sectional inertia radius r as

$$I = r^2 A = \rho^2 A^2 \tag{11.1}$$

with a non-dimensional description of r by $r = \rho\sqrt{A}$, i.e., $\rho = r/\sqrt{A}$.

Concept of equivalent length

The critical column buckling force N_C for different BC may with equivalent length be expressed generally

$$\lambda_C = -\pi^2 \;\Rightarrow\; N_C = \frac{-\pi^2 EI}{l_e^2} \tag{11.2}$$

© Springer International Publishing AG, part of Springer Nature 2018
S. L. Wiggers and P. Pedersen, *Structural Stability and Vibration*, Springer Tracts
in Mechanical Engineering, https://doi.org/10.1007/978-3-319-72721-9_11

As examples, the ratios for the equivalent length to the physical length of the five Euler cases in Table 3.1 of Chap. 3 are $l_e/L = 1, \simeq 0.7, 0.5, 2, 1$. For other problems, the ratio l_e/L might be estimated from an estimated buckling mode. Note that for vibration problem, the effective length normally decreases for higher order eigenmodes.

The column longitudinal stress σ_C, just before buckling, is assumed evenly distributed

$$\sigma_C = \frac{N_C}{A} = \frac{-\pi^2 EI}{l_e^2 A} = \frac{-\pi^2 E}{\Gamma^2} \quad \text{with} \quad \Gamma := \frac{l_e}{r} = \frac{l_e}{\rho\sqrt{A}} \tag{11.3}$$

defining the slenderness ratio Γ.

Concept of slenderness ratio

The critical column forces determined are only correct if $|\sigma_C|$ is not too large. A material yield stress σ_y is a limitation, and even for stresses below that, the actual material may have exceeded a proportionality limit σ_p that defines the limit for linear elasticity. The values σ_y and σ_p are positive values related to yielding and linear proportionality in compression. The N_C column force with a restriction on the stress just before buckling is therefore stated

Limiting stresses σ_y and σ_p

$$N_C = \frac{-\pi^2 EI}{l_e^2} \quad \text{for} \quad |\sigma_C| = \frac{\pi^2 E}{\Gamma^2} < \sigma_p \leq \sigma_y \tag{11.4}$$

11.2 Maximum Stress in Beam-Columns

The main conclusion of the beam-column theory is the nonlinear displacement dependence on the column force. From this follows directly, that strains and stresses also depend nonlinear on the column force and might give rice to large stresses. The intention of the present section is to localize and determine the numerical largest stress through the determination of the largest bending moment, still based on the assumption of linear elasticity. Influence from material nonlinearity and shear forces are treated in later sections of the present chapter.

In the outermost cross-sectional position (c from beam axis), the longitudinal stress σ from column force N and bending moment M is

$$\sigma = \frac{N}{A} + \frac{cM}{I} \quad \text{with} \quad M = M(\xi, N) \tag{11.5}$$

Largest longitudinal stress

Numerically largest stress $|\sigma|_{max}$ after determination of largest bending stress is

$$|\sigma|_{max} = \frac{|N| + \frac{c}{r^2}|M|_{max}}{A} \quad \text{with} \quad I = r^2 A \tag{11.6}$$

applying the cross-sectional radius of inertia r.

The largest bending moment depends on the column force N and on position ξ which in general also depends on N, i.e., $\xi = \xi(N)$. Determination of $|M|_{max}$ is therefore not simple. With reference to the solution (4.13) and (5.1), it is possible for a given beam-column problem to determine the displacement $y(\xi)$ and then the second derivative $y''(\xi)$. An extremal analysis of $M(\xi) = E I y''(\xi)/L^2$ determines $|M|_{max}$. A simple example illustrates this procedure.

Internal moment position

Figure 4.1 shows a beam-column for which specifically is assumed $M_1 = -M_0$. The resulting displacement function $y(\xi)$ for this case gives from (4.6)

$$y(\xi) = \frac{M_0 L^2}{\alpha^2 E I} \left(\cos(\alpha\xi) - \cot\alpha \sin(\alpha\xi) - 1 + \frac{\sin(\alpha\xi)}{\sin\alpha} \right)$$

$$y''(\xi) = \frac{M_0 L^2}{E I} \left(-\cos(\alpha\xi) + \cot\alpha \sin(\alpha\xi) - \frac{\sin(\alpha\xi)}{\sin\alpha} \right) \tag{11.7}$$

that also include the second derivative $y''(\xi)$ and thereby the extremal values of $M(\xi)$ by differentiation

$$\sin(\alpha\xi) + \cot\alpha \cos(\alpha\xi) - \frac{\cos(\alpha\xi)}{\sin\alpha} = 0 \quad \Rightarrow$$

$$\tan(\alpha\xi) = \frac{1 - \cos\alpha}{\sin\alpha} \quad \Rightarrow \quad \xi = 0.5 \tag{11.8}$$

For this, example $|M|_{max}$ is located independent of N. The result also follows directly from consideration of symmetry.

The actual bending moment then follows from $M(\xi) = E I y''(\xi = 0.5)$

$$|M|_{max} = |M_0| \left| -\cos\frac{\alpha}{2} + \cot\alpha \sin\frac{\alpha}{2} - \frac{\sin\frac{\alpha}{2}}{\sin\alpha} \right| = |M_0|\frac{1}{\cos\frac{\alpha}{2}} \geq |M_0|$$

$$\text{with} \quad -\lambda = \alpha^2 = \frac{-NL^2}{E I} = \frac{-N\Gamma^2}{E A} \tag{11.9}$$

applying trigonometric relations for simplification. Note that $|M|_{max} \to \infty$ for $\frac{\alpha}{2} \to \frac{\pi}{2}$, i.e., $\frac{NL^2}{E I} \to -\pi^2$.

Eccentric column force

For an analyzed example with $M_0 = Ne$ where e is eccentricity for position of column force N. The relations (11.6) with (11.9) give

$$|\sigma|_{max} = \frac{|N|}{A}\left(1 + \frac{ec}{r^2}\frac{1}{\cos\frac{\alpha}{2}}\right) \text{ or } \frac{|N|}{A} = \frac{|\sigma|_{max}}{\left(1 + \frac{ec}{r^2}\frac{1}{\cos\frac{\alpha}{2}}\right)}$$

$$\text{with } 0 < \alpha = \Gamma\sqrt{\frac{-N}{EA}} < \pi \text{ with } -N = P \tag{11.10}$$

Figure 11.1 shows $\frac{-N}{A} = \frac{P}{A}$ as a function of the slenderness Γ with $\frac{ec}{r^2}$ as a parameter. The curves are based on $E = 200\,\text{GPa}$ and $\sigma_{max} = \sigma_y = 250\,\text{MPa}$, as used in Fig. 11.23 in (Beer and Johnston 1992). A complete derivation of the problem is given in Sect. 11.5 in (Beer and Johnston 1992). Solutions to the implicit function of $\frac{P}{A}$ as a function of Γ are obtained by Newton–Raphson iterations.

Special beam-column load case

The treated problem is special in the sense that the beam load and the column load are increasing simultaneously. The influence from the parameter $\frac{ec}{r^2}$ acts as a weight factor to the influence from the column force through the factor $\frac{1}{\cos\frac{\alpha}{2}}$. For small values of α, $\cos\frac{\alpha}{2} \rightarrow 1$ and $\frac{|N|}{A} = \frac{|\sigma|_{max}}{1+\frac{ec}{r^2}}$. The formula (11.10) is (Beer and Johnston 1992) named the secant formula for eccentric column loading.

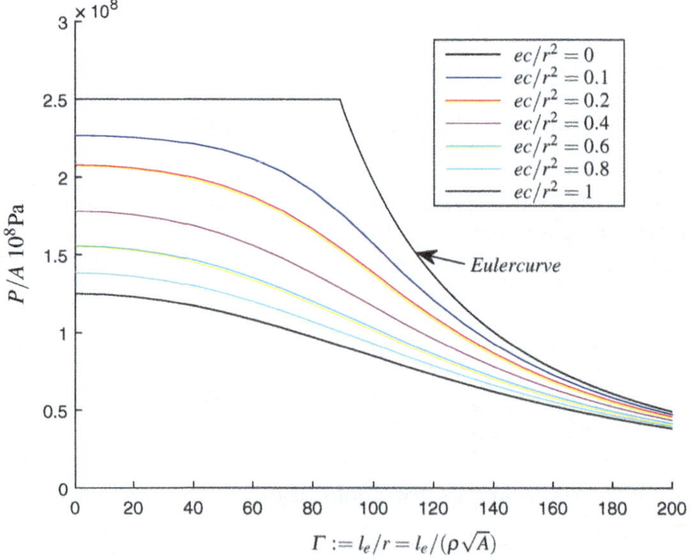

Fig. 11.1 Compressive force P per area A as a function of slenderness ratio Γ, with relative eccentricity ec/r^2 as parameter and material $\sigma_y = 250$ MPa $E = 200$ GPa

11.3 Material Nonlinearity

Before analysis of $|\sigma_C| > \sigma_p$, it might be useful to list a few moduli of elasticity such that it is seen in a broader perspective

- E is Young's modulus $= \frac{d\sigma}{d\epsilon}$ for $|\sigma| < \sigma_p$
- E_t is tangent modulus $= \frac{d\sigma}{d\epsilon}$ in general
- E_s is secant modulus $= \frac{\sigma}{\epsilon}$
- E_r is reduced modulus with $E_t < E_r < E$

Determination of buckling force with buckling stresses numerically larger than σ_p has an interesting historical evolution, which is shortly described. Finally, norms as the solution for design decisions are discussed.

11.3.1 Different Suggested Formulas

Euler 1744 and Engesser 1889

The result (11.2) is obtained by Euler in 1744 and first in 1889 Engesser suggested a formula also valid for $|\sigma_C| > \sigma_p$

$$N_C = \frac{-\pi^2 E_t(\sigma) I}{l_e^2} \quad \rightarrow \quad \sigma_C = \frac{-\pi^2 E_t(\sigma)}{\Gamma^2} \tag{11.11}$$

Note that (11.2) is explicit in contrast to (11.11) that is implicit. The critical column force must therefore be determined iteratively.

A few years after (1895), Engesser modified the result (11.11), taking into account that the cross-sectional parts are responding differently by start of buckling. From pure compressive state, bending courses further pressure in part of the cross section while tension in part of the cross section is less stressed. By unloading, the actual modulus is E and not E_t.

Karman 1910 and Shanley 1946

A reduced modulus of elasticity E_r that accounts for further loading by E_t and unloading by E is determined by Karman in 1910, and the formula (11.11) is modified to

$$N_C = \frac{-\pi^2 E_r(\sigma) I}{l_e^2} \quad \rightarrow \quad \sigma_C = \frac{-\pi^2 E_r(\sigma)}{\Gamma^2} \quad \text{where}$$

$$E_r = E_r(E, E_t, \text{cross-sectional shape}) \tag{11.12}$$

with detailed determination of E_r by Karman, based on perfect column, perfect load, and extreme slow loading. Careful experiments show results between the formulas (11.11) and (11.12). The explanation is given by Shanley in 1946. Even small imperfections result in bending displacements before the buckling force. Therefore, a smaller part of the cross section is unloaded as assumed by Karman. Imperfect systems are then more close to (11.11).

11.3.2   Experiments and Safety Factor

Figure 11.2 shows two curves for compressive stress related to beam buckling as a function of slenderness ratio Γ. The red curve represents the critical stress $|\sigma_C| = \pi^2 E/\Gamma^2$ until yield stress σ_y (or proportionality limit σ_p) for decreasing slenderness ratio is reached. For slenderness, ratio less than $\Gamma < \pi\sqrt{E/\sigma_y}$ constant critical stress is assumed $|\sigma_C| = \sigma_y$. The lowest slenderness for linear elasticity is $\Gamma_p < \pi\sqrt{E/\sigma_p}$.

The green curve with a dashed part represents the allowable compressive stress σ_A for which a safety factor μ is not specified. It is assumed that the safety factor has a constant value μ_E for $\Gamma \geq \pi\sqrt{E/\sigma_p}$, in the Fig. 11.2 chosen to 2. As the critical stress in the domain $0 < \Gamma < \pi\sqrt{E/\sigma_p}$ is uncertain, the safety factor is only available at its end values. The following assumptions may be applied for the allowable design stress σ_A in the range $0 \leq \Gamma \leq \Gamma_p$ and is determined by the end conditions of the interval

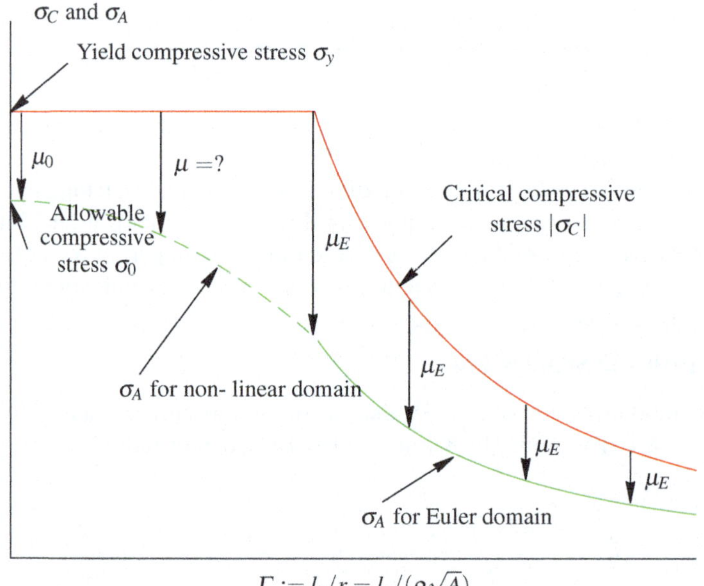

Fig. 11.2 Critical compressive stress$|\sigma_C|$ and allowable compressive stress σ_A as a function of slenderness ratio Γ. Factor of safety in the Euler range μ_E, but else not explicitly known

$$\sigma_A(\Gamma = 0) = \sigma_0 \quad \text{and} \qquad \sigma_A(\Gamma = \Gamma_p) = \frac{\sigma_p}{\mu_E}$$

$$\frac{d\sigma_A}{d\Gamma}(\Gamma = 0) = 0 \quad \text{and} \qquad \frac{d\sigma_A}{d\Gamma}(\Gamma = \Gamma_p) = -\frac{2\pi^2 E}{\mu_E \Gamma_p^3} \tag{11.13}$$

Different formulas to satisfy continuity and slope continuity at the joint point for the Euler domain and the nonlinear domain are suggested and used in norms and specifications. Reference is given to Sect. 11.6 in (Beer and Johnston 1992) that for practical procedures refer to (Steel 1989), to (Aluminum 1986), and to (Timber 1985).

The following design formulas are an applied procedure used in truss optimization.

11.3.3 Design Formulas Used in Truss Optimization

In research on optimal design of trusses, it is important to apply a detailed analytical model for compressive bars. The following procedure, directly related to Sect. 11.3.2, gives the minimum cross-sectional area A_{min} for a given compressive force $P > 0$.

Limiting compressive force

Two design domains are involved: the linear elastic domain also named the Euler domain and the nonlinear domain that without a safety factor may involve nonlinear elasticity or plasticity. These domains are separated by the compressive force P_p determined as

$$P_p = \frac{\sigma_p^2 l_e^2}{\pi^2 \rho^2 E \mu_E} \tag{11.14}$$

with effective length l_e, cross-sectional relation $\rho = \frac{\sqrt{I}}{A}$, Young's modulus E, proportionality limit stress σ_p and safety factor μ_E constant in the whole Euler domain. The limiting force P_p will result in $\sigma = \frac{\sigma_p}{\mu_E}$ with $A_{min} = \frac{P_p \mu_E}{\sigma_p}$.

Design in linear domain

Assumed that the actual compressive force is smaller than P_p, the design in the linear domain is

$$A_{min} = \frac{\sqrt{P} l_e}{\pi \rho \sqrt{\frac{E}{\mu_E}}} \quad \text{for } P \leq P_p \tag{11.15}$$

Design in nonlinear domain

Assumed that the actual compressive force is larger than P_p, the design in the nonlinear domain is

$$A_{min} = \frac{P - P_p}{\sigma_0} + \frac{\sqrt{P_p}\,l_e}{\pi\rho\sqrt{\frac{E}{\mu_E}}} \quad \text{for } P \geq P_p \qquad (11.16)$$

Note that a factor of safety in the nonlinear domain is not decided but is implicit involved in the allowable stresses σ_0 and $\frac{\sigma_p}{\mu_E}$ at the ends of the domain. Continuity in the two design formula is directly seen, but depending on the chosen values σ_0 and μ_E, a discontinuity in slope is possible. This follows from the choice of a parabolic function from σ_0.

Design for tensile forces

For tensile members, the minimum area is directly for the tensile force $N > 0$ by

$$A_{min} = \frac{N}{\sigma_A} \quad \text{for } N > 0 \ (P < 0) \qquad (11.17)$$

where the allowable stress is constant and mostly larger than σ_0, say by $\sigma_0 = 0.8\sigma_A$. For detailed optimal design of trusses, see (Pedersen 1973).

11.4 Improved Beam Modeling

In the initial Chap. 2, it is assumed that strains and displacements from shear forces are neglected. The important parameter to justify this is the slenderness ratio, as defined in (11.1) and the goal of the present section is to obtain a quantitative measure for the usefulness of simple beam theory.

Shear forces

The applied beam theory in the earlier chapters $\frac{d^2 y}{dx^2} = \frac{M}{EI}$ is based only on longitudinal stresses along the beam axis and therefore, in principle only valid for uniform moment load. With non-constant distribution of moments, the beam is also subjected to a shear force distribution $T = T(x)$ with

$$T(x) = -\frac{dM(x)}{dx} \qquad (11.18)$$

as static equilibrium in Fig. 11.3a shows.

Shear stresses

The distribution of shear stresses over the beam cross section, in equilibrium with a given shear force is complicated to determine as this is a 3D elastic problem. Good references to this problem is Cowper (1966, 1968). For the present discussion, the field of shear stresses τ are simply modeled by

$$\tau = \mu\frac{T}{A} \quad \text{with } 1 < \mu < 6 \qquad (11.19)$$

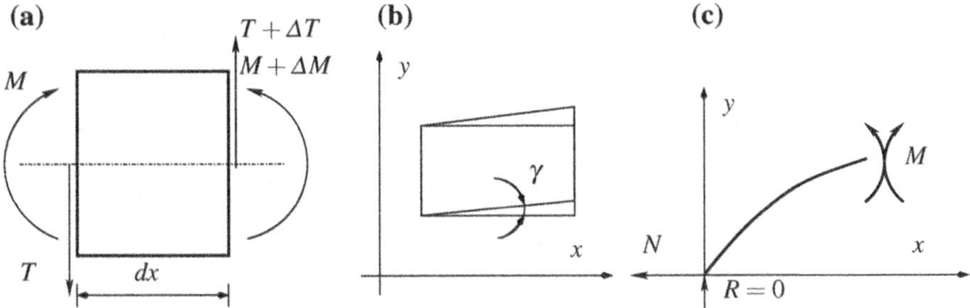

Fig. 11.3 **a** Relation between T and M, **b** definition of shear angle, **c** statics for moment equilibrium

where τ is the shear stress at the beam axes, A the cross-sectional area and μ a factor related to the distribution of shear stresses over the cross section. For solid double symmetric cross sections, a factor in the interval $1.1 < \mu < 1.5$ is often used.

Shear strains

For linear elastic material, the shear strain γ at the beam axis is

$$\gamma = \frac{\tau}{G} = \frac{2(1 + \nu)\tau}{E} \tag{11.20}$$

where ν is Poisson's ratio.

Slope of displacement

From (11.18), (11.19) and (11.20) it is then possible to determine the slope of beam displacement resulting from τ as seen in Fig. 11.3b

$$\left(\frac{dy}{dx}\right)_{\text{from } \tau} = \gamma = \frac{2(1 + \nu)\mu}{EA} \tag{11.21}$$

Improved beam model

The beam modeling when adding to (11.18) the shear stress displacements (by differentiation of (11.21)) is

$$\frac{d^2y}{dx^2} = \frac{M}{EI} - \frac{d^2M}{dx^2}\frac{2(1 + \nu)\mu}{EA} \tag{11.22}$$

11.4.1 *Influence on Column Buckling*

For the specific example of a simply supported column in Fig. 11.3c, the buckling column force is determined based on the improved beam model. The actual moment distribution is

$$M = Ny \quad \rightarrow \quad \frac{d^2M}{dx^2} = N\frac{d^2y}{dx^2} \tag{11.23}$$

With non-dimensional length (as applied earlier), $\xi = x/L$ and $y' = \frac{\partial y}{\partial \xi}$, then with (11.22) and (11.23)

$$y'' = \lambda y - \delta \lambda y'' \quad \text{with} \quad \lambda := \frac{NL^2}{EI} \quad \text{and}$$

$$\delta := \frac{2(1+v)\mu I}{AL^2} = \frac{2(1+v)\mu}{\Gamma^2} \quad \text{with} \quad \Gamma^2 := \frac{AL^2}{I} = \frac{L^2}{r^2} \tag{11.24}$$

where Γ is the slenderness ratio, that for $\Gamma = 100$ gives $\delta \simeq 10^{-3}$ and for $\Gamma = 20$ gives $\delta \simeq 10^{-2}$.

With the assumed BC the eigenvalue problem is

$$y'''' - \Lambda y'' = 0 \quad \text{with} \quad \Lambda := \frac{\lambda}{1 + \delta\lambda} \quad \text{and BC}$$

$$y(0) = y''(0) = y(1) = y''(1) = 0 \tag{11.25}$$

Quantitative errors for simple beam model

The solution to this problem is earlier determined to $\Lambda = -\pi^2$, and the buckling column force is determined as a function of δ

$$-\pi^2 = \frac{\lambda_C}{1 + \delta\lambda_C} \quad \rightarrow \quad \lambda_C = \frac{-\pi^2}{1 + \delta\pi^2} \tag{11.26}$$

The error without shear stresses is therefore 1% for $\delta \simeq 10^{-3}$ and 10% for $\delta \simeq 10^{-2}$.

11.4.2   Modeling Conclusions

- From a practical point of view then with slenderness ratio $\Gamma < 20$, an improved beam model should be applied.
- The performed analysis is for a simply supported column, but has general value if the effective length l_e is used for other boundary conditions.
- Note that the improved beam equation (11.22) has similar effects on static beam-column problems and on beam-column vibrations. For this later case, the effective length depends on the specific vibration mode.
- Applying numerical-based methods (finite element), it is normal to apply a Timoshenko beam model as (11.22) and not the more simple Bernoulli-Euler model. This is related to the fact that the time requirement for the numerical calculations more or less is the same.

Chapter 12
Large Displacements, Pre-buckling Strains, and Snap-Through

From material nonlinearity to influence from geometrical nonlinearity. For large displacements but still with linear strains, the solution to the Elastica is derived. The modified data for length and moment of inertia, just before buckling, also influence the determined buckling load. An example of a column, build up of truss bars, is presented with analytical result as well as a numerical 2D truss model. Linear elastic snap-through of von Mises truss is the final part of the chapter.

Large displacements and specifically large strains are complicated to analyze. However, in Chap. 14 Green–Lagrange strains are introduced. In this chapter, three specific cases are analyzed with reference mainly to the post-buckling behavior. A general large strain analysis is not included as this must be carried through by numerical models.

12.1 Large Displacements for the Elastica

In the present section, it is shown how large displacements may from a geometrical point of view require an improved beam model. A correct measure for curvature is applied, not the approximation $\frac{d^2 y}{dx^2}$ that are applied in the earlier sections.

Improved beam model

The experimental example in Fig. 7.1a shows in domain C that even with large displacements, they may be limited. The conflict between simple analysis and experiments is due to some assumed approximations. The present section shows that just a more accurate formula for curvature can describe essential aspects of an experiment.

© Springer International Publishing AG, part of Springer Nature 2018
S. L. Wiggers and P. Pedersen, *Structural Stability and Vibration*, Springer Tracts
in Mechanical Engineering, https://doi.org/10.1007/978-3-319-72721-9_12

Fig. 12.1 Cantilever column
with large displacements,
named the Elastica

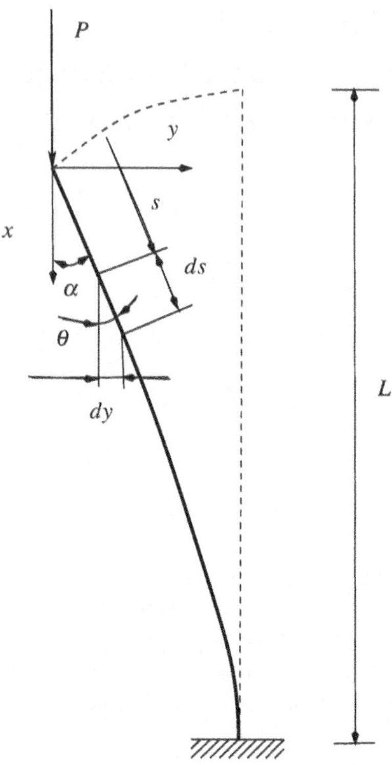

The Elastica

Figure 12.1 shows a uniform cantilever loaded with an end compressive force P with
a fixed direction (orthogonal to the fixed support). A state of large displacement is
analyzed. The improved beam model with κ being curvature is described by

$$M = \pm EI\kappa \tag{12.1}$$

where the sign $+$ or $-$ follows from the chosen signs for M and κ. Bending stiffness
EI is constant.

Expressed by the angle θ and the length s as alternative to y, x, the curvature κ is

$$\kappa = \frac{d\theta}{ds} \quad \text{and only for} \quad \frac{dy}{dx} \ll 1 \quad \rightarrow \quad \kappa \simeq \frac{d^2 y}{dx^2} \tag{12.2}$$

Static equilibrium with the external force P is therefore more correctly

$$EI\frac{d\theta}{ds} = -Py \tag{12.3}$$

that by differentiation with respect to s is

$$EI\frac{d^2\theta}{ds^2} = -P\frac{dy}{ds} = -P\sin\theta \tag{12.4}$$

This DE is analog to the equation for a vibrating pendulum, this is known as Kirchhoff's dynamic analogy.

Kirchhoff's dynamic analogy

By integration of (12.4)

$$\frac{1}{2}\left(\frac{d\theta}{ds}\right)^2 = \frac{P}{EI}\cos\theta + C \;\; \text{with a constant } C \tag{12.5}$$

as seen by differentiation of (12.5). The constant C is determined by the boundary condition. From $s = L, \theta = 0$ and $\frac{d\theta}{ds} = 0$ as $M = 0$ follows with $\theta = \alpha$ for $s = 0$

$$\left(\frac{d\theta}{ds}\right)^2 = 2(\cos\theta - \cos\alpha)\frac{P}{EI} \tag{12.6}$$

Since θ decreases for increasing s, then

$$ds = \frac{-d\theta}{\sqrt{2(\cos\theta - \cos\alpha)\frac{P}{EI}}} \tag{12.7}$$

that with neglected change of length L give

$$L = \int ds = \int_\alpha^0 \frac{-d\theta}{\sqrt{2(\cos\theta - \cos\alpha)\frac{P}{EI}}} = \sqrt{\frac{EI}{2P}} \int_0^\alpha \frac{d\theta}{\sqrt{\cos\theta - \cos\alpha}} \tag{12.8}$$

Inverse numerical solution

Mathematical transformation to an elliptical integral for which tables of functions are available is omitted, since an inverse solution is directly possible. So the alternative to find $\alpha = \alpha(P)$ is to determine $P = P(\alpha)$ which is done numerically explicitly from

$$P = \frac{EI}{2L^2}\left(\int_0^\alpha \frac{d\theta}{\sqrt{\cos\theta - \cos\alpha}}\right)^2 \tag{12.9}$$

Figure 12.2 shows curves for specific results with the end rotation $\alpha = \alpha(P)$. In reference (Timoshenko and Gere 1961) Sect. 2.7 page 76–82 further details and references for the Elastica are given. As examples, the relative force for $\alpha = 20^o, 60^o, 120^o$ is $\frac{P}{P_C} = 1.015, 1.152, 1.884$. The full curves in Fig. 12.2 are

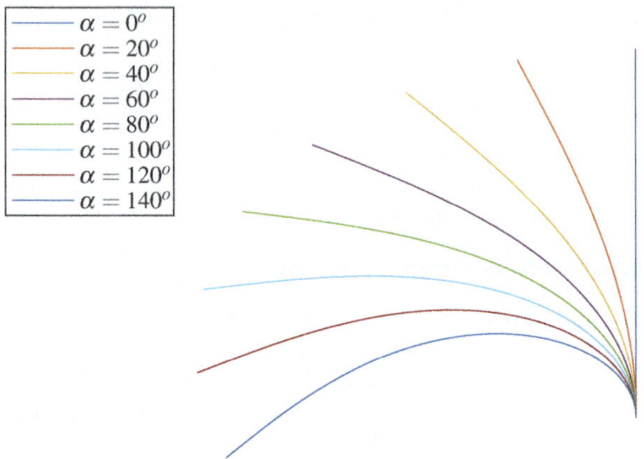

Fig. 12.2 Illustration of the solution to "The Elastica," by the inverse solution (12.9) for the end rotation

for the present book obtained by an extended inverse approach, i.e., the curves are obtained by finding $s(\theta)$ from (12.7). For each α-value θ is varied in the interval $[0, \alpha]$ having $s(\alpha) = 0$ and $s(0) = L$.

12.2 Pre-buckling Strain and Modified Data

The column buckling force N_C and the corresponding longitudinal strain ε_C in the beam axis just before buckling for the Euler columns are

$$N_C = \frac{-\pi^2 EI}{l_e^2} \;\Rightarrow\; \varepsilon_C = \frac{\sigma_C}{E} = \frac{-\pi^2}{\Gamma^2} \tag{12.10}$$

in terms of effective buckling length l_e and slenderness ratio $\Gamma = \frac{l_e}{r} = \frac{l_e}{\alpha\sqrt{A}}$ where the cross-sectional radius of inertia r is also expressed by a non-dimensional value α. In reality, the values of E, I, A, L and thereby Γ that are valid at the initiation of buckling should be applied and not the values in the non-loaded state.

Geometrical changes

While the material nonlinearities are treated in Chap. 11, geometrical nonlinearities are treated in this chapter. When a column in its length direction is loaded in compression, it gets shorter and due to Poisson's ratio also gets thicker which increases the cross-sectional area and moment of inertia. Results of all these changed data are a larger buckling compressive force but a smaller buckling stress.

Quantitative error measure

Mostly, the geometrical corrections are neglected and a quantitative measure for the errors as a function of slenderness is determined. For a solid cross-section, the following approximations are applied

$$(l_e + \Delta l_e)^2 = l_e^2 \left(1 + 2\frac{\Delta l_e}{l_e} + \left(\frac{\Delta l_e}{l_e}\right)^2 \right) \simeq l_e^2 (1 + 2\varepsilon_C)$$

$$I + \Delta I \simeq I(1 - 4\nu\varepsilon_C) \quad \text{with} \quad \varepsilon_C = \frac{-\pi^2}{\Gamma^2} \tag{12.11}$$

where ν is Poisson's ratio. Inserting (12.11) in (12.10), the column buckling force is

$$N_C = \frac{-\pi^2 EI}{l_e^2} \frac{1 - 4\nu\varepsilon_C}{1 + 2\varepsilon_C} \simeq \frac{-\pi^2 EI}{l_e^2}(1 - 2\varepsilon_C(1 + 2\nu)) \tag{12.12}$$

Taking $\nu = 0.3$ for a specific value, the buckling force is decreased $3.2\varepsilon_C \cdot 100\%$

Let the slenderness ratio be 20, then (12.10) gives $\varepsilon_C \simeq \frac{-10}{400} = -0.025$ and the error in neglecting the pre-buckling modifications is 8%, i.e., of the same order as not including the displacements from shear forces in Sect. 11.4. For slenderness ratio of 100, the error is only 0.3%. According to Chap. 11 then for $\Gamma < 100$ it is necessary to take the change in modulus of elasticity into account and therefore the here determined errors for solid cross-sections are relative small.

However, for non-solid columns, the axial compression before the material reaches its limit of proportionality in elasticity may be much larger, and the above treated modifications are necessary. A numerically analyzed truss column shows this.

12.2.1 A Build Up Column

A truss column, i.e., a column build as a truss structure, is numerically analyzed but also analyzed as a beam. This model can be found in (Timoshenko and Gere 1961) page 137. The one-dimensional beam analysis is compared to a 2D nonlinear truss analysis that illustrates the post-buckling results. Figure 12.3 shows with dotted lines for non-loaded state and full lines for displaced state corresponding to increased compressive force P.

For this truss structure, a formula is given in (Timoshenko and Gere 1961)

$$P_C = \frac{\pi^2 EI}{L^2} \frac{1}{1 + \frac{\pi^2 EI}{L^2} \frac{1}{2AE \sin\phi \cos^2\phi}} = 2.30 \cdot 10^6 \, \text{N} \tag{12.13}$$

accounting for the influence from shear deformation, but without geometrical modifications. The actual data are $E = 2 \cdot 10^{11}$ Pa, bar areas $A = 10^{-4} \, \text{m}^2$, the section width (of the six sections) of the truss column $b = 0.3$ m, the moment of inertia $I = 2A(\frac{b}{2})^2 = 4.5 \cdot 10^{-6} \, \text{m}^4$, the total length of the truss column $L = 1.8$ m and the angle of the cross bars is $\phi = 45^o$.

Note that, the displacements in Fig. 12.3 are correct size (1:1) and length and width changes in the truss structure are large. Modify the quantities in (12.13) according to

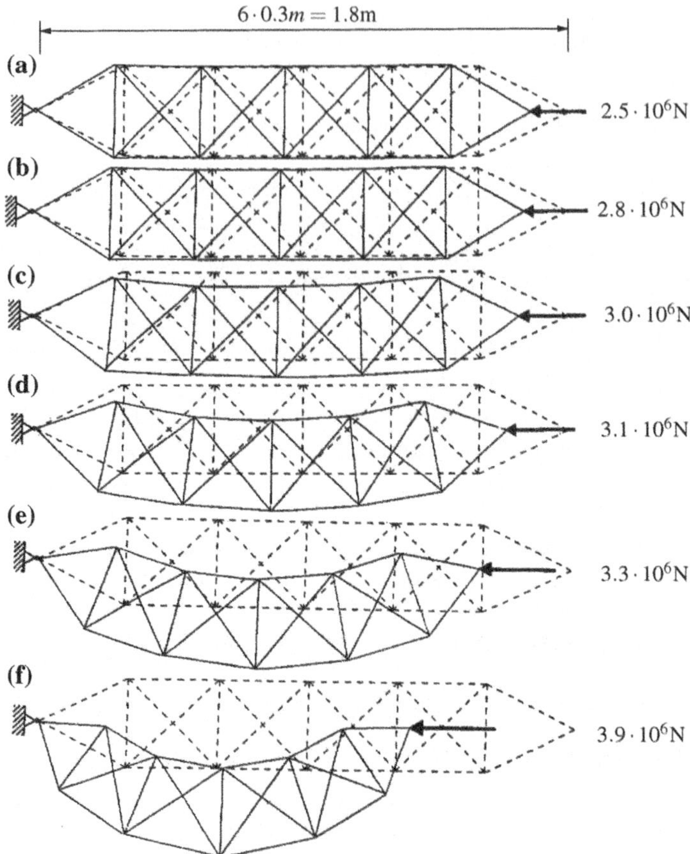

Fig. 12.3 Truss column displacements for increasing compressive end force P

$b = (0.3 + 0.01)\,\text{m}$, $L = (1.8 - 0.16)\,\text{m}$, and $\phi = 42.3^o$ giving the modified critical load

$$P_C = 2.85 \cdot 10^6 \,\text{N} \qquad (12.14)$$

which is an increase of 24%. Slenderness decreases from $\Gamma = 12 \to 10.5$ corresponding to an expected error of 22%.

The two-dimensional nonlinear truss analysis is iterative with collected results in Fig. 12.4 from which is read

$$P_C = 2.98 \cdot 10^6 \,\text{N} \qquad (12.15)$$

in practical agreement with the above.

Fig. 12.4 Collected truss column maximum displacements δ in transverse direction as a function of the compressive end force P, from Fig. 12.3

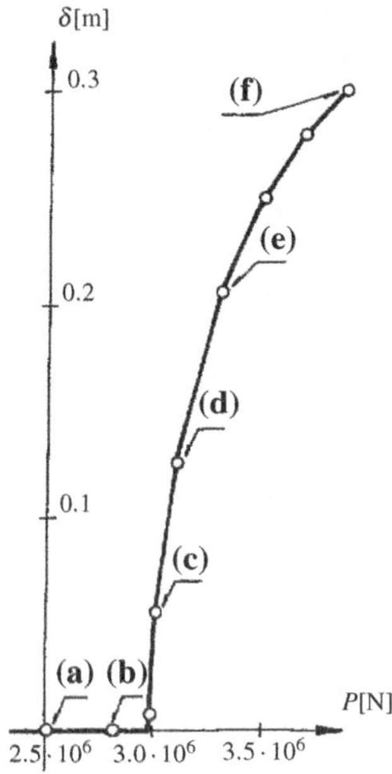

12.2.2 Notes on Global and Local Buckling and Vibrational Modes

The analyzed truss column calls for a short note on global and local modes, which is essential in both stability and vibration. For buildup structures as the truss column above (but also for stiffened plates and shells), there is a possibility for global buckling as the above analysis has assumed. However, a local bar may become critical resulting in local buckling, and in fact, it may be shown that the above example does this. The design procedure in Sect. 11.3.3 is a procedure accounting for local buckling and not global buckling.

Global and local buckling

For a statically non-determined structure, a local buckling may not result in structural collapse, often stated as fail-safe. The special case of simultaneous local and global buckling may also be the case, and the conclusion is that a structure should be analyzed for the possibility of all the cases. The present few comments also hold for vibrational analysis.

Global and local vibrational modes

Numerical eigenvalue solution for a specific structure or continuum is the tool that can solve stability and vibration problem, accounting for both local behavior and global behavior, and the present book should incorporate these formulations. However, it should be remembered that simpler more analytical oriented models give a basic understanding that is more difficult to obtain from specific numerical models.

12.3 Linear Elastic Snap-Through

The buckling force or the critical column force is in earlier chapters defined as the force where the formulation for an ideal column (homogeneous DE without beam load) shows arbitrary displacement. The formulation for a beam-column (non-homogeneous DE with beam load) shows singularities with large amplitudes. Figure 12.5a repeats the result in Fig. 7.1a, now with two other experimental results Fig. 12.5b, c. To analyze a model behind such results, it is necessary to perform a nonlinear post-critical analysis, as also performed in Sect. 10.5.

The von Mises truss

A classical simple example, named von Mises truss, illustrates the important aspects and is therefore analyzed in the present section. Figure 12.6 shows a two-bar truss with simple supports and two bars mutually connected by a charnier (no moment transfer). In addition, a figure for setting up the equilibrium is shown.

The geometrical relations, the physical relations, and the static equilibrium give

$$\text{Geometry:} \sin \widetilde{\alpha} = \frac{h - \delta}{\widetilde{L}}, \quad \widetilde{L} = \sqrt{(h - \delta)^2 + L^2 - h^2} = \sqrt{L^2 + \delta^2 - 2h\delta}$$

$$\text{Physics:} \quad L - \widetilde{L} = \delta L = \varepsilon L = \frac{\sigma L}{E} = \frac{SL}{EA}$$

$$\text{Statics:} \quad 2S \sin \widetilde{\alpha} = P \tag{12.16}$$

and thereby, the force P as a function of the displacement is

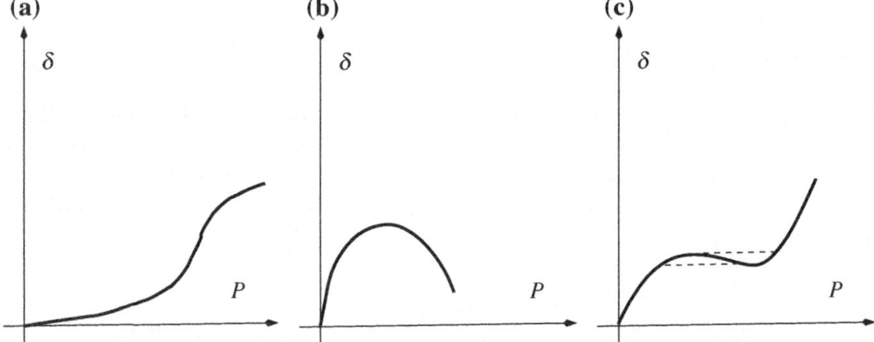

Fig. 12.5 Different experimental results, characteristic displacements δ as a function of compressive force P

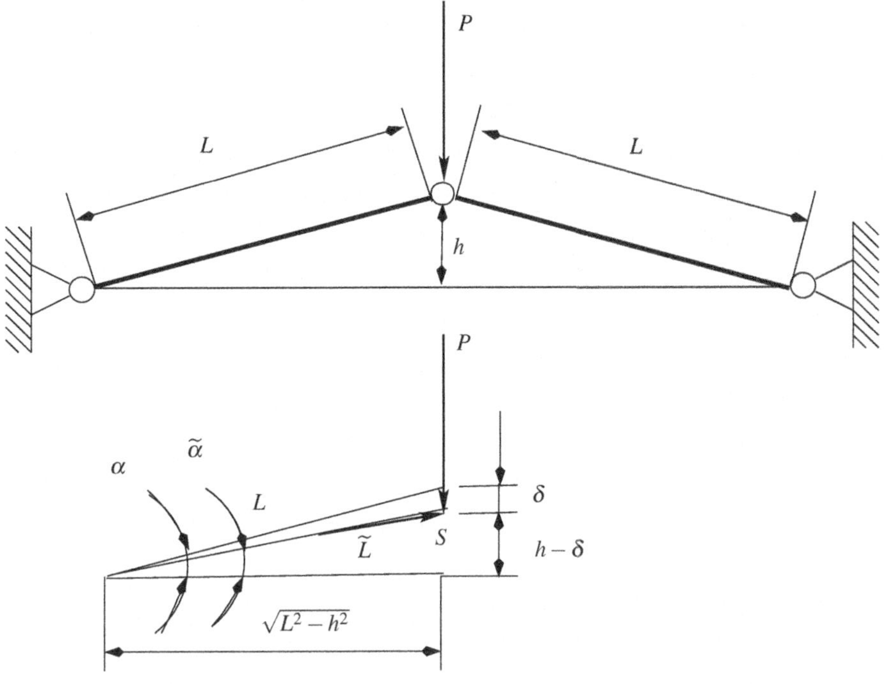

Fig. 12.6 Von Mises truss with definition of forces, lengths, and angles

$$P = \frac{2EA(h - \delta)}{L} \left(\frac{L}{\sqrt{L^2 + \delta^2 - 2h\delta}} - 1 \right) \quad \Rightarrow$$

$$\frac{P}{2EA} = \frac{h}{L} \left(1 - \frac{\delta}{h} \right) \left(\frac{1}{\sqrt{1 + (\frac{h}{L})^2 \frac{\delta}{h}(\frac{\delta}{h} - 2)}} - 1 \right) \tag{12.17}$$

Stable or unstable equilibrium

With $\frac{h}{L}$ as parameter, the function (12.17) is illustrated in Fig. 12.7, i.e., the dimensionless force $\frac{P}{2EA}$ as a function of dimensionless displacement $\frac{\delta}{h}$. Note that the solution is antisymmetric relative to $\delta = h$. Three states of equilibrium exist for $P = 0$, but $\delta = h$ is unstable, while $\delta = 0$ and $\delta = 2h$ are stable equilibriums. For this linear elastic problem, there is therefore possibility for different stable positions without plasticity.

The state of $\delta = 2\,h$ is reached by a load displacement pattern as for $\frac{h}{L} = 0.11$ and is indicated in Fig. 12.7 by the numbers 1 - 2 - 3 - 4 - 3 - 5. For unchanged force P, the snap-through from 2 to 3 is dynamic and here it is assumed that the inertia forces are damped before further load to position 4 and thereafter load decreases to 5.

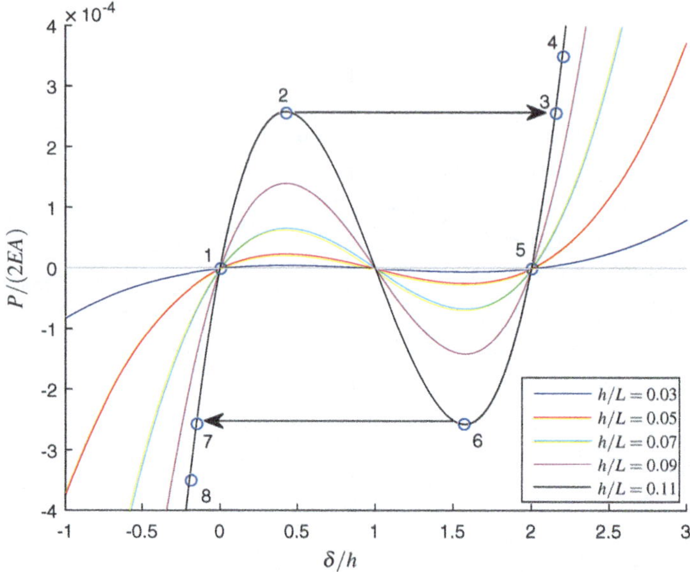

Fig. 12.7 Nonlinear function for relative force $\frac{P}{2EA}$ as a function of relative displacement $\frac{\delta}{L}$, with $\frac{h}{L}$ as parameter and quantities defined in Fig. 12.6

Snap-through mechanisms

Instability as the present snap may also be a positive aspect. Snap-through mechanisms are often used practically, as they may be designed to act reversible for a perfect elastic material. Alternative to the simple von Mises truss are cylinders and other shells that may result in elastic buckles that are easy to repair.

Coned disk springs = Belleville springs

The coned disk springs, named Belleville springs, are after many years (Belleville patent 1867) still an important machine element that may be designed to obtain rather different characteristics. The nonlinear stiffness function is in general well described by a classical (Almen and Laszlo 1936) analytical expression where the dependence on Poisson's ratio for the applied isotropic material is still discussed. Results from finite element (FE) axisymmetric models based on Green–Lagrange strains show only little influence from Poisson's ratio, and much confirm the analytical expression that also for a large extent has been confirmed by physical experiments. For FE analysis and design of coned disk springs, see (Pedersen and Pedersen 2011).

Chapter 13
Dynamics of Discretized Linear Systems

Discretization of continuous systems as by the finite element (FE) method is here restricted to linear systems. The main focus is on mass matrices, damping matrices and stiffness matrices and on a linear eigenvalue problems. Derived knowledge on real, positive eigenvalues, on mutual orthogonal eigenvalues and on mode expansion for general linear systems. Finally, vibrations with single degree of freedom (dof) give introductory information on the influence from damping.

13.1 Discretization Continuous to Discrete

Problems with a finite number of degree of freedom (dof) are often convenient for theoretical as well as for numerical analysis. Physical problems are continuous and therefore a discretization is necessary, i.e., a continuous model is approximated by a model with a finite number of dof.

Several well-established methods for such discretization exist

- Finite element (FE) method(s).
- Finite difference method(s).
- Ritz or Rayleigh-Ritz method.
- Weighted residual's methods such as
  Galerkin/Bubnov-Galerkin, Collocation, and least-squared methods.

Without discussing the individual specifics of these methods, it may be said that the finite element method is of major practical importance. Note that FE models have nodal dof, i.e., finite number of elements as well as finite number of dof. In the present chapter only linear elastic systems are analyzed. These systems may be solved numerically as described in Chap. 16, most effectively by the subspace iteration method.

© Springer International Publishing AG, part of Springer Nature 2018
S. L. Wiggers and P. Pedersen, *Structural Stability and Vibration*, Springer Tracts
in Mechanical Engineering, https://doi.org/10.1007/978-3-319-72721-9_13

13.2   Dynamic Equilibrium

For a linear system with finite number of dof, the dynamic equilibrium is

$$[M]\{\ddot{\widetilde{D}}\} + [C]\{\dot{\widetilde{D}}\} + [S]\{\widetilde{D}\} = \{\widetilde{A}\} \qquad (13.1)$$

indicating by $\widetilde{}$ (tilde) that displacements $\{\widetilde{D}\}$ and forces $\{\widetilde{A}\}$ are function of time t and a $\dot{}$ (dot) is notation for partial differentiation with respect to time t. The components in mass matrix $[M]$ are m_{ij}, defined as the inertia force for parameter i resulting from unit acceleration of parameter j. In analogy, the damping matrix $[C]$ is defined by damping forces c_{ij} resulting from unit velocity and for the stiffness matrix $[S]$ elastic forces s_{ij} from unit displacement.

Homogeneous solutions

The solution(s) for the homogeneous part of (13.1), i.e., for free vibrations with $\{\widetilde{A}\} = \{0\}$, is separated in an amplitude factor $\{\Delta_r\}$ and an exponential time function $e^{z_r t}$

$$\{\widetilde{D}\} = \{\Delta_r\}e^{z_r t} \ \text{ with } \ r = 1, 2, ..., n \qquad (13.2)$$

where z_r is complex

$$z_r = \alpha_r + i\omega_r \qquad (13.3)$$

and n is the order of the matrices; eigenmodes $\{\Delta_r\}$ describing time-independent amplitude with corresponding eigenvalue z_r satisfying

$$e^{z_r t}(z_r^2[M] + z_r[C] + [S])\{\Delta_r\} = \{0\} \qquad (13.4)$$

i.e., by the characteristic equation from the determinant condition, determine the eigenvalue(s)

$$|z_r^2[M] + z_r[C] + [S]| = 0 \qquad (13.5)$$

Instability condition

As a solution may be given by an arbitrary linear combination of solutions (13.2), then a positive real part ($\alpha_r > 0$) in (13.3) of just a single z_r implies instability.

13.3 Non-damped, Free Vibrations

As a basis reference system, non-damped ($[C] = [0]$) vibrations are of major importance. The eigenvalues are determined by

$$\left| z_r^2[M] + [S] \right| = 0 \tag{13.6}$$

with corresponding eigenmodes $\{\Delta_r\}$ evaluated from

$$\left(z_r^2[M] + [S] \right) \{\Delta_r\} = 0 \tag{13.7}$$

from which follows that the eigenmode $\{\Delta_r\}$ may be multiplied by an arbitrary constant. If the matrices $[M]$ and $[S]$ are both assumed symmetric and positive definite, then all z_r^2 are real and negative which imply ($\alpha = 0$, $\omega_r^2 > 0$) and all $\{\Delta_r\}$ are real and mutual orthogonal, all this to be proved below.

From (13.7) follows that a complex z_r implies that also $\{\Delta_r\}$ must be complex when the coefficient matrices only contain real values. Further, the complex conjugated eigen pairs must also be a solution as arbitrary analysis with complex quantities imply corresponding complex conjugated results.

13.3.1 Real, Positive Eigenfrequencies

Eigenfrequency proofs

Two scalar quantities are primarily defined as

$$\widehat{T}_{qr} := \{D_q\}^T [M]\{D_r\} \tag{13.8}$$

that may be termed mutual kinetic energy, and

$$\widehat{U}_{qr} := \{D_q\}^T [S]\{D_r\} \tag{13.9}$$

that may be termed mutual elastic energy. The $\widehat{}$ (hat) is added because the displacement vectors $\{D_q\}, \{D_r\}$ are not yet specified. As $\left(\{D_q\}^T [M]\{D_r\} \right)^T = \{D_q\}^T [M]^T \{D_r\}$, then symmetry of $[M]$ implies

$$\widehat{T}_{qr} := \widehat{T}_{rq} \tag{13.10}$$

and symmetry of $[S]$ implies

$$\widehat{U}_{qr} := \widehat{U}_{rq} \tag{13.11}$$

The matrices $[M]$ and $[S]$ are then also termed self-adjoint. Choosing specifically the eigenmodes $\{\Delta_g\}$, $\{\Delta_r\}$, the hat index $\widehat{\ }$ is removed and from (13.7) follows by multiplication with $\{\Delta_g\}^T$

$$z_r^2 T_{qr} + U_{qr} = 0 \tag{13.12}$$

and analogously

$$z_q^2 T_{rq} + U_{rq} = 0 \tag{13.13}$$

that with (13.10) and (13.11) also valid without the hat index $\widehat{\ }$ give

$$(z_q^2 - z_r^2) T_{qr} = 0 \tag{13.14}$$

Assuming z_r^2 complex, it is proved that this cannot be the case as the conjugated solution $\bar{z}_r^2$, $\{\bar{\Delta}_r\}$ with $\bar{\ }$ (bar) index is also a solution that is substituted for the q solution in (13.14)

$$(\bar{z}_r^2 - z_r^2) T_{\bar{r}r} = 0 \tag{13.15}$$

and if $T_{\bar{r}r} \neq 0$ the eigenvalues can only be real.

Real eigenvalues

By $[M]$ positive definite it is then proved that the eigenvalues are real.

$$T_{\bar{r}r} = \left(\Re\{\Delta_r\}^T - \Im\{\Delta_r\}^T\right)[M]\left(\Re\{\Delta_r\}^T + \Im\{\Delta_r\}^T\right) > 0 \tag{13.16}$$

Positive ω^2 eigenvalues

A necessary and sufficient condition for $z_r^2 < 0\,(\omega_r^2 > 0)$ is a positive definite stiffness matrix $[S]$

$$\widehat{U}_{qr} = \{D_q\}^T[S]\{D_r\} > 0 \tag{13.17}$$

for all admissible $\{D_r\}$.

13.3.2　Orthogonal Eigenmodes

Eigenmodes proof

Assuming different values of eigenvalues $z_q \neq z_r$ for $q \neq r$, this gives by (13.14) with (13.8) inserted

$$T_{qr} = \{\Delta_q\}^T [M]\{\Delta_r\} = \delta_{qr} T_r \tag{13.18}$$

where δ_{qr} is Kronecker's delta and T_r may by scaling of $\{\Delta_q\} = \{\Delta_r\}$ be normalized to $T_r = 1$.

Mutual orthogonal eigenmodes

The result of (13.18) expresses that eigenmodes are mutual orthogonal (weighted with the mass matrix), and from (13.18) and (13.12) follows

$$U_{qr} = \{\Delta_q\}^T [S]\{\Delta_r\} = -\delta_{qr} \omega^2 T_r \tag{13.19}$$

13.4 Expansion for More General Linear Systems

Using the obtained results for non-damped free vibration, the more general dynamic equilibrium (13.1) is analyzed. This is also known as modal analysis. Assuming all solutions ω_r^2, $\{\Delta_r\}$ for corresponding non-damped, free system available, then the mutual orthogonal eigenmodes $\{\Delta_r\}$ are a complete system and therefore by linear combination any N dimensional vector $\{D\}$ can be described by

$$\{D\} = \sum_r \phi_r \{\Delta_r\} \tag{13.20}$$

where ϕ_r are time-dependent linear combination factors. Inserting (13.20) in (13.1) gives

$$[M] \sum_r \ddot{\phi}_r \{\Delta_r\} + [C] \sum_r \dot{\phi}_r \{\Delta_r\} + [S] \sum_r \phi_r \{\Delta_r\} = \{\widetilde{A}\} \tag{13.21}$$

that for a specific damping matrix leads to decoupled equations for each linear combination factor ϕ_r. A specific damping matrix is a linear combination of mass and stiffness matrix

$$[C] = \eta_r [M] + \tau_r [S] \tag{13.22}$$

which may have different combination factors for each mode r. This damping assumption is named Rayleigh damping.

Decoupled dynamic equations

Multiplying (13.21) with $\{\Delta_q\}$ an using (13.18), (13.20) and (13.22) results in

$$\ddot{\phi}_r + (\eta_r + \tau_r \omega^2)\dot{\phi}_r + \omega^2 \phi_r = \{\Delta_q\}^T \{\widetilde{A}\}/T_r \text{ for } r = 1, 2, ..., n \tag{13.23}$$

These one-dimensional problems are solved as shown in the section below. Relevant questions are: How realistic is the assumption on damping (13.22)? and how demanding is it to obtain all eigen pairs ω_r^2, $\{\Delta_r\}$? The assumption (13.22) includes three specific cases: (1) damping proportional to mass (volume) $[C] = \eta[M]$, (2) damping proportional to stiffness $[C] = \tau[S]$, and (3) and damping proportional to critical damping $[C] = \eta_r[M]$.

13.5   Vibrations with Single dof

A system with only a single dof is presented more simple as

$$m\ddot{\tilde{d}} + c\dot{\tilde{d}} + s\tilde{d} = \tilde{a} \tag{13.24}$$

as an alternative to (13.23). Solutions are found of the type

$$\tilde{d} = d e^{\alpha t} \sin(\omega t - \theta) \tag{13.25}$$

i.e., a harmonic vibration $\sin(\omega t - \theta)$ with time-independent part of the amplitude d and damping factor $e^{\alpha t}$. From (13.24) follows

$$\dot{\tilde{d}} = d e^{\alpha t} \left(\omega \cos(\omega t - \theta) + \alpha \sin(\omega t - \theta)\right) \tag{13.26}$$

$$\ddot{\tilde{d}} = d e^{\alpha t} \left((\alpha^2 - \omega^2) \sin(\omega t - \theta) + 2\alpha\omega \cos(\omega t - \theta)\right) \tag{13.27}$$

Inserting (13.25)–(13.27) in (13.24) gives

$$m d e^{\alpha t} \left(((\alpha^2 - \omega^2) + \frac{c}{m}\alpha + \frac{s}{m}) \sin(\omega t - \theta) + \omega(2\alpha + \frac{c}{m}) \cos(\omega t - \theta)\right) = \tilde{a} \tag{13.28}$$

from which a number of classical conclusions are drawn.

13.5.1   Non-damped Single dof Vibrations

For non-damped $c = 0$ and free $\tilde{a} = 0$ vibrations follow $\alpha = 0$ and θ as well as d are arbitrary. The squared non-damped eigenfrequency ω_1^2 is

$$\omega_1^2 = \frac{s}{m} \tag{13.29}$$

The particular solutions with a harmonic force

$$\tilde{a} = a \sin \Omega t \tag{13.30}$$

imply $\alpha = 0$, $\theta = 0$, and $\omega = \Omega$, i.e.,

$$d = \frac{a}{m} \Big/ \left(\frac{s}{m} - \Omega^2 \right) = \frac{a}{m} \Big/ \left(\omega_1^2 - \Omega^2 \right) \tag{13.31}$$

The ratio $\frac{d}{a}$ is named the amplification factor.

13.5.2 Damped Single dof Vibrations

For damped $c \neq 0$ but free $\tilde{a} = 0$ vibrations

$$\alpha = -\frac{c}{2m} \tag{13.32}$$

Damped free vibrations

When the factor $e^{\alpha t}$ is decreasing for $c > 0$, i.e., stable vibrations, but increasing for $c < 0$, i.e., unstable vibrations. The damped eigenfrequency is

$$\omega^2 = \frac{s}{m} - \frac{c^2}{4m^2} = \omega_1^2 - \frac{c^2}{4m^2} \tag{13.33}$$

Critical damping

The critical damping c_C is defined in relation to $\omega^2 = 0$, i.e.,

$$c_C^2 = 4sm = 4\omega_1^2 m^2 \tag{13.34}$$

Damped forced vibrations

With harmonic force (13.30), the solutions $\alpha = 0$, $\omega = \Omega$ imply d, θ determined by

$$md \left(\frac{s}{m} - \Omega^2 \right) \sin(\Omega t - \theta) + md\Omega \frac{c}{m} \cos(\Omega t - \theta) = a \sin \Omega t \tag{13.35}$$

with solution

$$d = \frac{a}{m} \Big/ \sqrt{\left(\frac{s}{m} - \Omega^2 \right)^2 + \left(\frac{c}{m}\Omega \right)^2}$$

$$\theta = \arctan \left(\frac{\frac{c}{m}\Omega}{\frac{s}{m} - \Omega^2} \right) \tag{13.36}$$

Written with the relative quantities $\frac{\Omega}{\omega}$ (frequency ratio) and $\frac{c}{c_C}$ (damping ratio), then equation (13.36) simplifies to

$$d = \frac{a}{s} / \sqrt{\left(1 - (\frac{\Omega}{\omega})^2\right)^2 + \left(2\frac{c}{c_C}\frac{\Omega}{\omega}\right)^2}$$

$$\theta = \arctan\left(\frac{\left(2\frac{c}{c_C}\frac{\Omega}{\omega}\right)}{1 - (\frac{\Omega}{\omega})^2}\right) \tag{13.37}$$

Note that d is always limited and that θ for $\frac{\Omega}{\omega} = 1$, i.e., resonance is $90°$ independent of $\frac{c}{c_C}$.

Chapter 14
Discretized Stability Analysis

Discretized stability analysis for continua is described for two different versions, named linear elastic buckling analysis and nonlinear elastic buckling analysis. These approaches include eigenvalue problems with different formulations. Linear buckling analysis is often non-reliable and therefore nonlinear buckling analysis, based on Green-Lagrange strains, are described in more detail. The chapter gives from a mathematical point of view a picture of this approach. Research results are still published, and the present chapter should be updated when new clarifying results are available. It is here assumed that the reader has some knowledge on geometrical nonlinear elasticity.

14.1 Different Stiffness Matrices

Nonlinear elastic = reversible

The static equations of equilibrium for an elastic problem, including reversible nonlinear elastic problems, is for a model with a finite number of degrees of freedom (dof) written

$$[S]\{D\} = \{A\} \tag{14.1}$$

where more general aspects of the stiffness matrix $[S]$ are discussed in detail. The displacement vector $\{D\}$ and the corresponding force vector $\{A\}$ relate in finite element (FE) analysis to a finite number of nodal degrees of freedom and (14.1) express equilibrium for a total model (system) or for a single element as primarily assumed.

© Springer International Publishing AG, part of Springer Nature 2018

S. L. Wiggers and P. Pedersen, *Structural Stability and Vibration*, Springer Tracts in Mechanical Engineering, https://doi.org/10.1007/978-3-319-72721-9_14

Influence parameters

For each element, the stiffness matrix depends on:

- the element geometry (form),
- the element orientation (and in axisymmetric problems also position),
- the element displacement assumption and nodal positions,
- the material parameters for the element,
- and for nonlinear models the actual stress/strain state.

Different stiffness matrices

For nonlinear models, different stiffness matrices are involved, such as

- $[S_0]$, symmetric stiffness matrix for a non-displaced model,
- $[S_s]$, non-symmetric secant stiffness matrix for a displaced model,
- $[S_\gamma]$, symmetric displacement gradient stiffness matrix for a displaced model,
- $[S_\sigma]$, symmetric stress stiffness matrix for a displaced model, and
- $[S_t] = [S_\gamma] + [S_\sigma]$, symmetric tangent stiffness matrix for a displaced model.

These different stiffness matrices are more specifically defined in Sect. 14.2 below. After this rather theoretical section, follows a description of linear bifurcation buckling which is more directly comprehended.

14.2   Green-Lagrange Strain Model

For a single strain component η_{ij} of the Green-Lagrange strain tensor, we have in matrix notation

$$\eta_{ij} := \{B_{ij}^0\}^T \{D\} + \frac{1}{2}\{D\}^T [B_{ij}^2]\{D\} \tag{14.2}$$

where $\{D\}$ contains all the element nodal displacements and the strain/displacement vector $\{B_{ij}^0\}$ (operator) gives the linear terms. The symmetric strain/displacement matrix $[B_{ij}^2]$ gives the nonlinear terms. For a specific element $\{B_{ij}^0\}$ and $[B_{ij}^2]$ are constant and do not depend on $\{D\}$.

With a variation of only $\{D\}$, we get directly from (14.2) with $[B_{ij}^2]$ symmetric

$$d\eta_{ij} = \{B_{ij}^0\}^T\{dD\} + \{D\}^T[B_{ij}^2]\{dD\} = \{B_{ij}^0\}^T\{dD\} + \{B_{ij}^L\}^T\{dD\} \tag{14.3}$$

where the defined vector $\{B_{ij}^L\}^T := \{D\}^T[B_{ij}^2]$ depends linearly on $\{D\}$.

Secant relation

Collecting all the strain components of the element, we can in strain vector notation write (14.2) and (14.3) as

$$\{\eta\} = \left([B^0] + \frac{1}{2}[B^L] \right) \{D\} = [B_s]\{D\} \tag{14.4}$$

Tangential relation

$$\{d\eta\} = ([B^0] + [B^L])\{dD\} = [B_t]\{dD\} \tag{14.5}$$

with the matrix $[B^L]$ linearly dependent on the nodal displacement $\{D\}$ while the matrix $[B^0]$ is independent of $\{D\}$. The index s in the matrix $[B_s]$ indicates that this matrix describes the secant relations (a kind of mean values), while the matrix $[B_t]$ describes the tangent relations. We also use the index s for the secant constitutive matrices and for the secant stiffness matrix.

14.2.1 *Finite Element Equilibrium and the Secant Stiffness Matrix*

The general equilibrium follows from the principle of virtual work ($\int \{\delta\eta\}^T \{\tau\}dV = \{\delta D\}^T \{A\}$)

$$\int [B_t]^T \{\tau\}dV = \{A\} \tag{14.6}$$

where the resulting stresses $\{\tau\}$ are conjugated to the differential strains $\{d\eta\}$ and $\{A\}$ are given nodal loads. These stresses we write in terms of resulting strains and further the resulting strains in terms of resulting displacements from (14.4)

$$\{\tau\} = [L_s]\{\eta\} = [L_s][B_s]\{D\} \tag{14.7}$$

with the constitutive secant relations described by the secant constitutive matrix $[L_s]$.

Non-symmetric secant stiffness matrix

Inserting (14.7) in (14.6), we get

$$\int [B_t]^T [L_s][B_s]dV\{D\} = [S_s]\{D\} = \{A\} \tag{14.8}$$

In general, the element secant stiffness matrix $[S_s] = \int [B_t]^T [L_s][B_s]dV$ is non-symmetric because $[B_t] \neq [B_s]$. However, in a Newton–Raphson approach to find a solution to (14.8) we do not need the system secant stiffness matrix, only the element secant stiffness matrices. The element secant stiffness matrices $[S_s]$ are the physically most important matrices because they determine the equilibrium. Note that the involved constitutive matrix $[L_s]$ is the secant constitutive matrix. Note that in FE literature secant matrices are seldom discussed.

14.2.2 Residuals and Newton–Raphson Iterations

Iterative procedure

An element residual $\{R\}$ is defined from (14.8)

$$\{R\} = [S_s]\{D\} - \{A\} = \sum_e ([S_s]_e\{D\}_e) - \{A\} \tag{14.9}$$

and we iteratively update our estimate $\{D\}$ by $\{\Delta D\}$ found from

$$\{\mathbf{R}\} + \{\Delta\mathbf{R}\} = \{\mathbf{0}\} \quad\Rightarrow\quad \{\Delta\mathbf{R}\} = [\mathbf{S}_t]\{\Delta\mathbf{D}\} = -\{\mathbf{R}\} \tag{14.10}$$

where the system (in bold) tangent stiffness matrix $[\mathbf{S}_t]$ is symmetric. This update is done on the system level, so that the system tangential stiffness matrix and the system residual vector must be assembled in the usual finite element manner.

To see that each element tangent stiffness matrix $[S_t]$ is symmetric, we calculate the differential $\{dR\}$ directly from (14.6) (assuming here $\{A\}$ not depending on $\{D\}$). We get two terms

$$d([S_s]\{D\}) = \int [dB_t]^T\{\tau\}dV + \int [B_t]^T\{d\tau\}dV \tag{14.11}$$

The symmetry related to the second term follows from $\{d\tau\} = [L]\{d\eta\}$ and (14.5)

$$[B_t]^T\{d\tau\} = [B_t]^T[L]\{d\eta\} = [B_t]^T[L_t][B_t]\{dD\} \tag{14.12}$$

and as the tangent constitutive relation $[L_t]$ is symmetric, so is $[B_t]^T[L_t][B_t]$.

14.2.3 The Stress Stiffness Matrix

Proof of symmetry

The first term on the right-hand side of (14.11) is named the stress stiffness matrix $[S_\sigma]$, and a proof of the symmetry of this matrix can be found, e.g., in (Crisfield 1991,1997). Here, we base the proof directly on the symmetry of the matrices $[B_{ij}^2]$

for the individual strain components, but we need some rearrangements. With $[B^0]$ independent of $\{D\}$ we have from (14.5)

$$[dB_t] = [dB^L] = \begin{bmatrix} \{dD\}^T[B_{11}^2] \\ \vdots \\ \{dD\}^T[B_{ij}^2] \\ \vdots \end{bmatrix} \tag{14.13}$$

and thus with $[B_{ij}^2]^T = [B_{ij}^2]$ (follows from the definition (14.2)) the transposed matrix is

$$[dB_t]^T = \left[[B_{11}^2]\{dD\}, \dots, [B_{ij}^2]\{dD\}, \dots \right] \tag{14.14}$$

Post-multiplying $[dB_t]^T$ with $\{\tau\}$ as needed in (14.11), we get with summation notation

$$[dB_t]^T\{\tau\} = \left(\sum \tau_{ij}[B_{ij}^2] \right)\{dD\} \tag{14.15}$$

with summation over all the stresses in the vector $\{\tau\}$.

14.2.4 The Tangent Stiffness Matrix for a Tetrahedron Element

As stated initially, each $[B_{ij}^2]$ is symmetric and thus the stress stiffness matrix is symmetric. Collecting the terms from (14.12) and (14.15), we have finally the element tangent stiffness matrix, that for a tetrahedron 3D element is

$$[S_t] = [S_\gamma] + [S_\sigma] = \int [B_t]^T[L_t][B_t]dV +$$

$$\int \left(\tau_{xx}[B_{xx}^2] + \tau_{yy}[B_{yy}^2] + \tau_{zz}[B_{zz}^2] + \tau_{xy}[B_{xy}^2] + \tau_{xz}[B_{xz}^2] + \tau_{yz}[B_{yz}^2] \right) dV \tag{14.16}$$

Note that the involved constitutive matrix $[L_t]$ in the first integral is the tangent constitutive matrix ($\{d\tau\} = [L]\{d\eta\}$), while for expressing the stresses τ_{ij} in the second integral in terms of the strains or displacement gradients we need the secant constitutive matrix $[L_s]$ (or a sequential update of stresses).

14.3   Linear Bifurcation Buckling

With the tangential stiffness matrix available for a FE model, buckling may be determined as described in (Cook et al. 2002), Chap. 18. For three most simple elements, see (Pedersen 2005a, b, 2006a), that give details of the theory described in Sect. 14.2. In bifurcation buckling as in Chap. 3 two close equilibrium states are possible for the same load. Before bifurcation linear as well as nonlinear elastic behavior is possible but the stress stiffness matrix is assumed linear depending on the load and is here related to a reference state $\{A\}_{ref}$.

$$[S_\sigma] = \lambda[S_\sigma]_{ref} \ \text{with} \ [S_\sigma]_{ref} \ \text{from}$$
$$[S_0]\{D\}_{ref} = \{A\}_{ref} \ \Rightarrow \ \{D\}_{ref} \ \Rightarrow \ \sigma_{ref} \ \Rightarrow \ [S_\sigma]_{ref} \tag{14.17}$$

With the linear assumption and unchanged conventional stiffness matrix $[S_0]$, two close equilibrium states are

$$\left([S_0] + \lambda[S_\sigma]_{ref}\right)\{D\}_{ref} = \lambda\{A\}_{ref}$$
$$\left([S_0] + \lambda[S_\sigma]_{ref}\right)\left(\{D\}_{ref} + \{\delta D\}\right) = \lambda\{A\}_{ref} \tag{14.18}$$

where $\{\delta D\}$ is the bifurcation mode from $\{D\}_{ref}$. The difference of these two equations gives an eigenvalue problem

$$\left([S_0] + \lambda[S_\sigma]_{ref}\right)\{\delta D\} = \{0\} \tag{14.19}$$

that may be solved, just like the eigenvalue problem (13.7) for non-damped, free vibrations. As for the vibrational eigenmodes, the critical stability mode $\{\delta D\}$ may be scaled and thereby the problem normalized. The problem (14.19) is identical with the dynamic problem (13.7) and theoretical results on real λ_C and mutual orthogonality of eigenmodes hold. However, only the lowest eigenvalue λ_C is of practical interest. Note that the stress stiffness matrix $[S_\sigma]$ is indefinite.

The critical load $\{A\}_C$ corresponding to the bifurcation displacement $\{\delta D\}$ is

$$\{A\}_C = \lambda_C\{A\}_{ref} \tag{14.20}$$

From the assumption of linearity between $\{A\}$ and $[S_\sigma]$ follows directly, that the critical buckling load vector $\{A\}_C$ is independent of the size(norm) of $\{A\}$. This implies uncertainty in linear buckling analysis. The relations between the individual stress components in a finite element are unchanged for linear buckling analysis. However, with geometrical nonlinear displacement analysis this is not the case, even assuming material linear elasticity. This also gives doubts to the estimated buckling load, obtained by linear buckling analysis.

14.3.1 Nonlinear Obtained Reference State

The procedure (14.17)–(14.20) may be based on a primarily solved nonlinear elastic equilibrium, iteratively obtained as in (14.10). Such solution is given the index n to indicate a step of a possible incremental load $\{A\}_n = \sum_{\tilde{n}=1}^{\tilde{n}=n} \{\varDelta A\}_{\tilde{n}}$.

The determined solution gives a determined stress stiffness matrix $[S_\sigma]$ and a displacement gradient stiffness matrix $[S_\gamma]$ by

$$[S_s]_n\{D\}_n = \{A\}_n \;\Rightarrow\; \{D\}_n \;\Rightarrow\; \gamma_n \;\Rightarrow\; [S_\gamma]_n \text{ and } \sigma_n \;\Rightarrow\; [S_\sigma]_n \quad (14.21)$$

and then modifying (14.18)–(14.20) for the next load step with λ_n for the extrapolated step give $\lambda = 1 + \lambda_n$ for buckling load determination

$$\big(([S_\gamma]_n + [S_\sigma]_n) + \lambda_n[S_\sigma]_n\big)\{\varDelta D\}_{n+1} = \big([S_\gamma]_n + (1 + \lambda_n)[S_\sigma]_n\big)\{\varDelta D\}_{n+1} =$$
$$\big([S_\gamma]_n + \lambda[S_\sigma]_n\big)\{\varDelta D\}_{n+1} = \lambda\{\varDelta A\}_{n+1}$$
$$\big([S_\gamma]_n + \lambda[S_\sigma]_n\big)\big(\{\varDelta D\}_{n+1} + \{\varDelta\}\big) = \lambda\{\varDelta A\}_{n+1} \quad (14.22)$$

where $\{\varDelta\}$ is the bifurcation mode from $\{\varDelta D\}_{n+1}$. The difference of the last two equations gives an eigenvalue problem

$$\big([S_\gamma]_n + \lambda_C[S_\sigma]_n\big)\{\varDelta\} = \{0\} \;\Rightarrow\; \text{ the eigen pair } \lambda_C, \{\varDelta\} \quad (14.23)$$

The critical load $\{A\}_C$ corresponding to the bifurcation displacement $\{\varDelta\}$ is

$$\{A\}_C = \lambda_C\{A\}_n \quad (14.24)$$

14.4 Nonlinear Implicit Procedure or Nonlinear Explicit Procedure

Two different procedures for incremental, iterative solution of the nonlinear problem may be applied.

Implicit procedure

The first procedure is named implicit and during the Newton–Raphson iterations (14.10) the tangent stiffness matrix is redefined at each iteration step and the next load increment is only taken after convergence test. This may allow for larger incremental load steps, but may be rather computational demanding. The nonlinear implicit method based on iteratively obtained secant stiffness is for a 1D problem illustrated in Fig. 14.1.

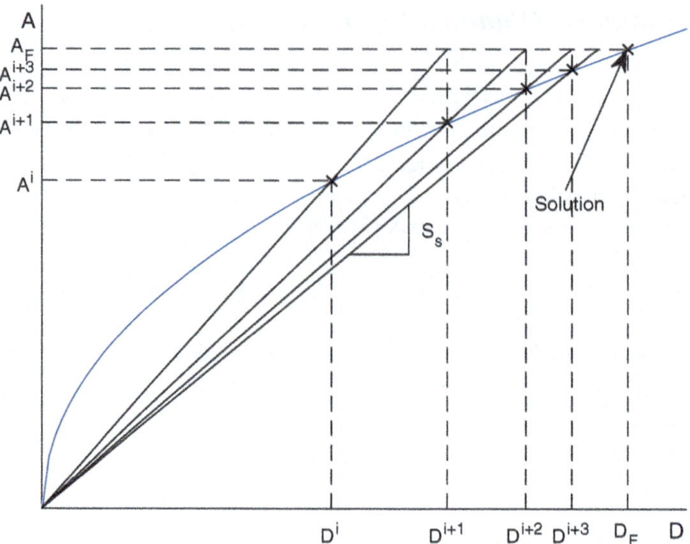

Fig. 14.1 Illustration of procedure to determine in 1D the displacement D_F for a given force A_F in a number of iterations with redefined secant stiffness matrix

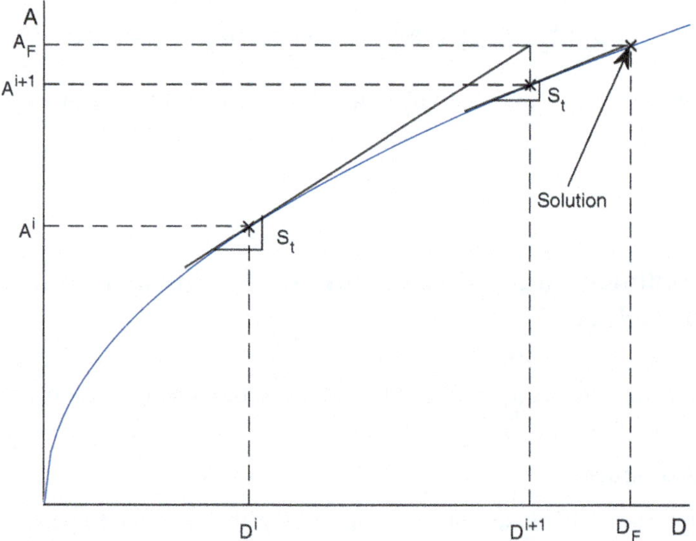

Fig. 14.2 Illustration of procedure to determine in 1D the displacement D_F for increment force up to A_F in a number of iterations with redefined tangent stiffness matrix

Explicit procedure

The second procedure for incremental, iterative solution is named explicit and only the tangent stiffness matrix corresponding to the state before the beginning of the actual load increment is needed. This means that for each load increment only one tangential stiffness matrix needs to be set up and "inverted" (see Chap. 16). However, many small load increments are necessary, and convergence must be checked by decreasing the size of these load increments. The nonlinear explicit/incremental method based on obtained tangent stiffness is for a 1D problem illustrated in Fig. 14.2.

Different combinations of implicit and explicit procedures are possible.

14.5   Dilatation Errors from Linear Strain Modeling

The difference between linear strain modeling and Green-Lagrange strain modeling can be large, and very sensitive to the boundary conditions. This dependence on actual boundary conditions is illustrated in Fig. 14.3 (from Pedersen 2005b). The plate is modeled by 8000 triangular (almost equal sized) ring elements giving in total 8242 degrees of freedom. Through the thickness we have 10 quadrangles each divided into

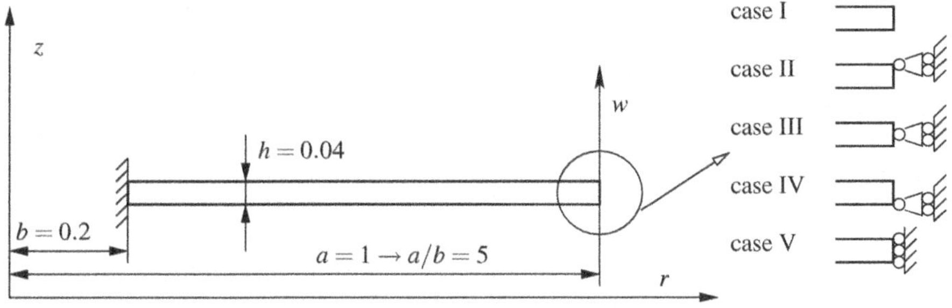

FE linear strains	FE non-linear strains	Linear strain error	Boundary condition case
$w = 0.044$ m	$w = 0.036$ m	+ 22 %	case I
$w = 0.025$ m	$w = 0.018$ m	+ 39 %	case II
$w = 0.044$ m	$w = 0.026$ m	+ 69 %	case III
$w = 0.025$ m	$w = 0.036$ m	- 31 %	case IV
$w = 0.018$ m	$w = 0.016$ m	+ 13 %	case V

Fig. 14.3 Axisymmetric model of a circular plate with a circular hole with five cases of boundary condition at the outer edge. Uniform pressure load $p = 10^6$ Pa in the z-direction, isotropic material with modulus of elasticity $E = 2 \cdot 10^{11}$ Pa and Poisson's ratio 0.3. All lengths in meter. Resulting displacement in axial direction at the outer edge is w. The symbolism in the case V indicates that this cross section cannot rotate. The size of the displacements is not large but only equals the thickness of the plate.

four triangles, and along the radial direction 200 quadrangles each divided into four triangles.

Rigid body rotations are not possible with linear strain models, because this will contradict the linearity of the results. Thus, the expected strain error (often stated in textbooks) is compensated by displacement errors (expansive dilatation) that depending on the boundary condition gives the background for understanding the final errors. The final errors are indirect, they can be positive as well as negative, and are due to an erroneous displacement field for problems which involve rotations, that need not be large. If an element is rotated the angle θ, then an erroneous pure expansive dilatation that amounts to $(1 - \cos\theta)$ is the result for linear strain models. To understand the negative error of -31% in case IV, we note that a compressive reaction at the lower corner (from erroneous expansion) will result in a moment that counteracts the moment distribution from the external pressure, and thus for this case result in a too small displacement w.

Chapter 15
Routh–Hurwitz-Liénard–Chipart Criteria

The description of the Routh–Hurwitz criteria for stability of linear, free   vibrations is presented in a number of text books, but seldom with more detailed analytical expressions. Furthermore, the simplifications by Liénard–Chipart (1914) are often not included. This is the background for the present chapter and for applying the name Routh–Hurwitz-Liénard–Chipart criteria.

15.1   Relation to Dynamic Stability

The discretized problem of linear, free vibrations is written

$$[M]\{\ddot{\tilde{D}}\} + [C]\{\dot{\tilde{D}}\} + [S]\{\tilde{D}\} = 0 \tag{15.1}$$

Assuming the separation

$$\{\tilde{D}\} = \{D\}e^{zt} = \{D\}e^{(\alpha+i\omega)t} \tag{15.2}$$

where $\{D\}$ is independent of time t, we get

$$e^{zt}\left([M]z^2 + [C]z + [S]\right)\{D\} = \{0\} \tag{15.3}$$

and thus the condition for a non-trivial solution $\{D\} \neq \{0\}$ is

$$\left|[M]z^2 + [C]z + [S]\right| = P(z) = 0 \tag{15.4}$$

© Springer International Publishing AG, part of Springer Nature 2018
S. L. Wiggers and P. Pedersen, *Structural Stability and Vibration*, Springer Tracts
in Mechanical Engineering, https://doi.org/10.1007/978-3-319-72721-9_15

i.e., our matrix problem is converted into a polynomial problem

$$P(z) = a_0 z^n + a_1 z^{n-1} + \cdots + a_{n-1} z + a_n = 0 \tag{15.5}$$

where the coefficients a_i are given by the elements of the square matrices $[M]$, $[C]$, $[S]$ and the order n is equal to twice the order of these matrices.

From a qualitative point of view, the important question is as follows: do all the roots z for $i = 1, 2, \ldots, n$ of polynomial $P(z)$ have negative real parts, i.e., $\alpha_i < 0$? If this is the case, it follows from (15.2) that the solution to (15.1) is asymptotically stable, i.e., stable and attractive.

The answer to the question just raised is given by the necessary and sufficient criteria of Routh–Hurwitz-Liénard–Chipart. The critical case of some $\alpha_i = 0$ will not be discussed here.

15.2   Necessary Criterion

The general proofs for these important criteria are too extensive to be given here. The proof by Routh (1955) first published in 1875 is based on the theorems of Sturm and the index of Cauchy; the proof by Hurwitz (1895) is related to some earlier work by Hermite; the proof by Liénard and Chipart (1914) is based on quadratic forms and gives some valuable simplifications of the Routh–Hurwitz criteria. For more extensive information, see Gantmacher (1959) or Hahn (1967), Popov (1962), where related geometrical criteria by Leonhard-Mikhailov are presented.

An important necessary criterion, which is most valuable from a practical point of view, is easily proved, and thus we shall do this. It states that all the coefficients a_i for $i = 0, 1, \ldots, n$ of the polynomial must necessarily be positive (or all negative by changing sign of (15.5)).

Let all the unknown roots of $P(z)$ have a negative real part

$$
\begin{aligned}
z_i &= -\alpha_i && \text{(real root)} \\
z_j &= -\alpha_j \pm i\omega_j && \text{(complex conjugated roots)} \\
&\text{for all } \alpha > 0 &&
\end{aligned}
\tag{15.6}
$$

then by normal factorization of the polynomial $P(z)$, we get the expansion in products of the real and the complex terms

$$
\begin{aligned}
P(z) &= \Pi_i(z - z_i) \\
&= \Pi_i(z + \alpha_i)\Pi_j(z + \alpha_j - i\omega_j)\Pi_j(z + \alpha_j + i\omega_j) \\
&= \Pi_i(z + \alpha_i)\Pi_j(z^2 + \alpha_j^2 + 2z\alpha_j + \omega_j^2)
\end{aligned}
\tag{15.7}
$$

from which it is directly seen that all terms are positive and that the resulting a_i for $i = 0, 1, 2, \ldots, n$ corresponding to this expanded specific case of (15.5) are thus all positive and real. Reading this in the inverse sense, we have the theorem:

A necessary condition for a polynomial with real coefficients to have roots all with negative real parts is that all the coefficients have the same sign.

Polynomials with complex coefficients, which are often encountered in dynamic stability, will be discussed later. The subject of dynamic stability is introduced in Chap. 9.

15.3 Necessary and Sufficient Criteria

The necessary and sufficient criteria are mostly expressed by the determinant conditions

$$H_i > 0 \quad \text{for } i = 1, 2, \ldots, n \tag{15.8}$$

where the Hurwitz determinants up to order eight are defined as left-upper sub-determinants

$$H_8 := \begin{vmatrix} a_1 & a_3 & a_5 & a_7 & & & & \\ a_0 & a_2 & a_4 & a_6 & a_8 & & & \\ & a_1 & a_3 & a_5 & a_7 & & & \\ & a_0 & a_2 & a_4 & a_6 & a_8 & & \\ & & a_1 & a_3 & a_5 & a_7 & & \\ & & a_0 & a_2 & a_4 & a_6 & a_8 & \\ & & & a_1 & a_3 & a_5 & a_7 & \\ & & & a_0 & a_2 & a_4 & a_6 & a_8 \end{vmatrix} \tag{15.9}$$

to be read in the sense that H_i is the determinant of order i defined in the upper left-hand corner. More specifically,

$$\begin{aligned} H_1 &:= a_1 \\ H_2 &:= a_1 a_2 - a_0 a_3 \\ H_3 &:= H_2 a_3 - (a_1 a_4 - a_0 a_5) a_1 \\ & \qquad \cdot \\ & \qquad \cdot \end{aligned} \tag{15.10}$$

It is also important to note that the last determinant involved is given by

$$H_n = H_{n-1} a_n \tag{15.11}$$

As mentioned, proof of (15.8) will not be given. Further, by means of the criteria of Liénard–Chipart, it is shown that the criteria (15.8) are not independent. In fact, if the odd-order determinants are positive, then those of even order are also positive and vice verse. Thus, the practical statement of necessary and sufficient criteria for negative real parts of the roots is

$$a_i > 0 \quad \text{for} \quad i = 0, 1, 2, \ldots, n$$
$$\text{and} \quad H_i > 0 \quad \text{for} \quad i = 2, 4, \ldots, n \quad \text{or} \quad (n - 1 \text{ with } n \text{ odd})$$
$$\text{or} \quad H_i > 0 \quad \text{for} \quad i = 3, 5, \ldots, n - 1 \quad \text{or} \quad (n \text{ with } n \text{ odd}) \qquad (15.12)$$

We see that the condition $H_1 > 0$ is omitted as $H_1 = a_1 > 0$. The simplified version (15.12) of the Routh–Hurwitz criteria (15.9) is what we term the Routh–Hurwitz-Liénard–Chipart criteria.

15.4   Criteria Expressed by Polynomial Coefficients

Although we are mainly concerned with problems described by a polynomial of even order n, we shall here list the specific criteria for the $n = 1, 2, 3, 4, 5, 6, 7, 8$, expressed by the polynomial coefficients a_i. Note that

$$a_i = 0 \quad \text{for} \quad i > n \qquad (15.13)$$

which simplifies the lower order criteria.

For all the following results, we have chosen the even Hurwitz determinants when n is odd and the odd Hurwitz determinants when n is even. Furthermore, relation (15.11) is encountered.

Even with these many simplifications, the expressions corresponding to n = 6, 7, and 8 are rather lengthy and we therefore define the following sub-determinants of order two

$$A_1 := a_1 a_2 - a_0 a_3 \qquad A_2 := a_1 a_4 - a_0 a_5$$
$$A_3 := a_1 a_6 - a_0 a_7 \qquad A_4 := a_3 a_4 - a_2 a_5$$
$$A_5 := a_3 a_6 - a_2 a_7 \qquad A_6 = a_5 a_6 - a_4 a_7 \qquad (15.14)$$

Then the final Routh–Hurwitz-Liénard–Chipart criteria are written as follows:

$n = 1:$ $a_0 z + a_1$

    Root with negative real part if and only if $a_i > 0$ for $i = 0, 1$

$n = 2:$ $a_0 z^2 + a_1 z + a_2$

    All roots with negative real part if and only if $a_i > 0$ for $i = 0, 1, 2$

$n = 3:$ $a_0 z^3 + a_1 z^2 + a_2 z + a_3$

    All roots with negative real part if and only if $a_i > 0$ for $i = 0, 1, 2, 3$ and

$$H_2 = A_1 = a_1 a_2 - a_0 a_3 > 0$$

$n = 4:$ $a_0 z^4 + a_1 z^3 + a_2 z^2 + a_3 z + a_4$

    All roots with negative real part if and only if $a_i > 0$ for $i = 0, 1, \ldots, 4$ and

$$H_3 = A_1 a_3 - a_1^2 a_4 = (a_1 a_2 - a_0 a_3) a_3 - a_1^2 a_4 > 0$$

$n = 5:$ $a_0 z^5 + a_1 z^4 + a_2 z^3 + a_3 z_2 + a_4 z + a_5$

    All roots with negative real part if and only if $a_i > 0$ for $i = 0, 1, \ldots, 5$ and

$$H_2 = A_1 = a_1 a_2 - a_0 a_3 > 0 \text{ and}$$

$$H_4 = A_1 A_4 - A_2^2 = (a_1 a_2 - a_0 a_3)(a_3 a_4 - a_2 a_5) - (a_1 a_4 - a_0 a_5)^2 > 0$$

$n = 6:$ $a_0 z^6 + a_1 z^5 + a_2 z^4 + a_3 z_3 + a_4 z^2 + a_5 z + a_6$

    All roots with negative real part if and only if $a_i > 0$ for $i = 0, 1, \ldots, 6$ and

$$H_3 = A_1 a_3 - A_2 a_1 > 0 \text{ and}$$

$$H_5 = A_1(A_4 a_5 + (2a_1 a_5 - a_3^2) a_6) + A_2(a_1 a_3 a_6 - A_2 a_5) - a_1^3 a_6^2 > 0$$

$n = 7:$ $a_0 z^7 + a_1 z^6 + a_2 z^5 + a_3 z_4 + a_4 z^3 + a_5 z^2 + a_6 z + a_7$

    All roots with negative real part if and only if $a_i > 0$ for $i = 0, 1, \ldots, 7$ and

$$H_2 = A_1 > 0 \text{ and}$$

$$H_4 = A_1(A_3 + A_4) - A_2^2 > 0 \text{ and}$$

$$H_6 = A_1(2A_3 A_6 + A_4 A_6 - A_5^2)$$
$$+ A_2(-A_2 A_6 + A_3 A_5) - A_3^2 > 0$$

$n = 8:$ $a_0 z^8 + a_1 z^7 + a_2 z^6 + a_3 z_5 + a_4 z^4 + a_5 z^3 + a_6 z^2 + a_7 z + a_8$

    All roots with negative real part if and only if $a_i > 0$ for $i = 0, 1, \ldots, 8$ and

$$H_3 = A_1 a_3 - A_2 a_1 > 0 \text{ and}$$

$$H_5 = A_1(A_3 a_5 + A_4 a_5 - A_5 a_3 + A_6 a_1 - a_1 a_3 a_8)$$
$$+ A_2(-A_2 a_5 + A_3 a_3 + a_1^2 a_8) - A_3^2 a_1 > 0 \text{ and}$$

$$H_7 = A_1(A_1 a_7^2 a_8 - A_2 a_5 a_7 a_8 + A_3(2A_6 a_7 - a_5 a_8 + a_3 a_7 a_8)$$
$$+ A_4(A_6 a_7 - a_5^2 a_8 + 2a_3 a_7 a_8) + A_5(-A_5 a_7 + a_3 a_5 a_8 - a_1 a_7 a_8))$$
$$+ A_6(-a_1 a_5 a_8) + a_8^2(-a_3^3 + 3a_1 a_3 a_5 - 3a_1^2 a_7))$$
$$+ A_2(A_2(-A_6 a_7 + a_5^2 a_8 - a_3 a_7 a_8) + A_3(A_5 a_7 + 3a_1 a_7 a_8 - a_3 a_5 a_8)$$
$$+ A_4(-a_1 a_7 a_8) - 2a_1^2 a_5 a_8^2 + a_1 a_3^2 a_8^2)$$
$$+ A_3(-A_3^2 a_7 + A_3 a_1 a_5 a_8 - a_1^2 a_3 a_8^2) + a_1^4 a_8^3 > 0 \tag{15.15}$$

The criteria for $n = 4, 5, 6$ are stated in Popov (1962) in the same form, but derived by the Mikhailov geometrical criteria.

15.5   Polynomials with Complex Coefficients

The criteria presented are only valid in relation to polynomials with real coefficients. It is, however, possible to transform a nth order polynomial with complex coefficients into a $2n$th order polynomial with real coefficients. By this transformation, which we shall perform below, the use of the Routh–Hurwitz-Liénard–Chipart criteria is extended to these general polynomials. Let us assume the polynomial to be

$$P(z) = c_0 z^n + c_1 z^{n-1} + \cdots + c_{n-1} z + c_n$$
$$\text{with }\ c_i = a_i + i b_i \ \text{ for }\ i = 0, 1, \ldots, n \tag{15.16}$$

where all a_i, b_i are real. The polynomial $\widetilde{P}(z)$, where all the coefficients of $P(z)$ are changed into the complex conjugated ones

$$\widetilde{P}(z) = \bar{c}_0 z^n + \bar{c}_1 z^{n-1} + \cdots + \bar{c}_{n-1} z + \bar{c}_n$$
$$\text{with }\ \bar{c}_i = a_i - i b_i \ \text{ for }\ i = 0, 1, \ldots, n \tag{15.17}$$

has the complex conjugated roots of $P(z)$. Now let us write

$$P(z) = Q(z) + i R(z) \ \text{ with }\ Q(z) = \Re(P(z)) \ \text{ and }\ R(z) = \Im(P(z))$$
$$Q(z) = a_0 z^n + a_1 z^{n-1} + \cdots + a_{n-1} z + a_n$$
$$R(z) = b_0 z^n + b_1 z^{n-1} + \cdots + b_{n-1} z + b_n \tag{15.18}$$

then the polynomial $\widetilde{P}(z)$ is written

$$\widetilde{P}(z) = Q(z) - i R(z) \tag{15.19}$$

and the product known to have all the roots of $P(z)$ as well as of $\widetilde{P}(z)$ is

$$P(z)\widetilde{P}(z) = Q^2(z) + R^2(z) \tag{15.20}$$

We see from the definitions in (15.18) that the polynomial $P(z)\widetilde{P}(z)$ has only real coefficients and is of order $2n$. The expressions for the real coefficients of $P(z)\widetilde{P}(z)$ are given below for $n = 1, 2, 3, 4$ such that the criteria of (15.15) for $n = 2, 4, 6, 8$ can be directly applied.

Using the definitions (15.18) and (15.20), we get

$n = 1$: $(a_0^2 + ib_0)z + (a_1 + ib_1)$

is transformed into

$$(a_0^2 + b_0^2)z^4 + 2(a_0a_1 + b_0b_1)z + (a_1^2 + b_1^2)$$

$n = 2$: $(a_0 + ib_0)z^2 + (a_1 + ib_1)z + (a_2 + ib_2)$

is transformed into

$$(a_0^2 + b_0^2)z^4 + 2(a_0a_1 + b_0b_1)z^3 + (2(a_0a_2 + b_0b_2) +$$
$$(a_1^2 + b_1^2))z^2 + 2(a_1a_2 + b_1b_2))z + (a_2^2 + b_2^2)$$

$n = 3$: $(a_0 + ib_0)z^3 + (a_1 + ib_1)z^2 + (a_2 + ib_2)z + (a_3 + ib_3)$

is transformed into

$$(a_0^2 + b_0^2)z^6 + 2(a_0a_1 + b_0b_1)z^5 + (2(a_0a_2 + b_0b_2) + a_1^2 + b_1^2)z^4 +$$
$$2(a_0a_3 + b_0b_3 + a_1a_2 + b_1b_2)z^3 + (2(a_1a_3 + b_1b_3) + a_2^2 + b_2^2)z^2 +$$
$$2(a_2a_3 + b_2b_3)z + (a_3^2 + b_3^2)$$

$n = 4$: $(a_0 + ib_0)z^4 + (a_1 + ib_1)z^3 + (a_2 + ib_2)z^2 + 2(a_2a_3 + b_2b_3)z + (a_4 + ib_4)$

is transformed into

$$(a_0^2 + b_0^2)z^8 + 2(a_0a_1 + b_0b_1)z^7 + (2(a_1a_3 + b_1b_2) + a_1^2 + b_1^2)z^6 +$$
$$2(a_03 + b_0b_3 + a_1a_2 + b_1b_2)z^5 + (2(a_0a_4 + b_0b_4) + a_1a_3 + b_1b_3) + a_2^2 + b_2^2)z^4 +$$
$$2(a_1a_4 + b_1b_4 + a_2a_3 + b_2b_3)z^3 + (2(a_2a_4 + b_2b_4) + a_3^2 + b_3^2)z^2 +$$
$$2(a_3a_4 + b_3b_4)z + (a_4^2 + b_4^2) \tag{15.21}$$

15.6 Example with Complex Coefficients

For the problem of a vibrating column, the time-dependent part of the solution is governed by the equation

$$\ddot{\Psi}(\tau) + \beta\dot{\Psi}(\tau) + (\zeta \pm i\varepsilon)\Psi = 0 \tag{15.22}$$

where β is a real damping parameter and $(\zeta \pm i\varepsilon)$ is the eigenvalue resulting from the spatial eigenvalue problem. Inserting the exponential solution

$$\Psi(\tau) = e^{zt} \tag{15.23}$$

we get the polynomial condition

$$z^2 + \beta z + (\zeta \pm i\varepsilon) = 0 \tag{15.24}$$

By means of the transformation (15.21) for $n = 2$, we get with $a_0 = 1$, $b_0 = 0$, $a_1 = \beta$, $b_1 = 0$, $a_2 = \zeta$, $b_2 = \pm\varepsilon$,

$$z^4 + 2\beta z^3 + (2\zeta) + \beta^2)z^2 + 2\beta\zeta z + (\zeta^2 + \varepsilon^2) = 0 \qquad (15.25)$$

The stability criteria for this is given by (15.15) for $n = 4$ and the result is

$\beta > 0, \zeta > 0$ and

$$(2\beta(2\zeta + \beta^2) - 2\beta\zeta)2\beta\zeta - (2\beta^2)^2(\zeta^2 + \varepsilon^2) = 4\beta^2(\beta^2\zeta - \varepsilon^2) > 0 \;\;\rightarrow\;\; \frac{\varepsilon^2}{\zeta} < \beta^2 \quad (15.26)$$

in agreement with Pedersen (1977b), where a direct analysis of this problem is performed. An example of a third- order equation with complex constants is given by Morgan and Sinha (1983).

Chapter 16
Numerical Tools

A number of numerical methods are important tools for stability and vibration analysis. They are here described on the introductory level, but even a few FORTRAN codes for solving linear equations are included. These few code lines often cover the most time (CPU time)-consuming part for a large program. The mathematical description of subspace iteration is given special attention.

16.1 Newton–Raphson Iterations for Solving Nonlinear Problems

Method for a scalar function

The method named Newton–Raphson is a general method for solving a nonlinear problem like $f(x) = 0$. Let first f and x be scalar quantities, and a guess at the ith iteration x_i returns the error (residual) $f(x_i) \neq 0$. Then, using the first-order derivative $\nabla f = \frac{df}{dx}$, the searched value is updated by

$$f(x_i) + \nabla f_i \cdot \Delta x_i = 0 \quad \Rightarrow \quad x_{i+1} = x_i + \Delta x_i \tag{16.1}$$

If the procedure converges, it normally converges in few iterations. Note that $\nabla f = 0$, or from a numerical point of view close to zero, must be specifically taken care of.

Method for a vector function

When f and x are vectors $\{f\}$ and $\{x\}$, a gradient matrix $[\nabla f_k]$ must be inverted, and a multivariable Newton–Raphson iteration is

$$\{f(x_i)\} + [\nabla f_i]\{\Delta x_i\} = 0 \quad \Rightarrow \quad \{x_{i+1}\} = \{x_i\} - [\nabla f_i]^{-1}\{f(x_i)\} \tag{16.2}$$

© Springer International Publishing AG, part of Springer Nature 2018

S. L. Wiggers and P. Pedersen, *Structural Stability and Vibration*, Springer Tracts in Mechanical Engineering, https://doi.org/10.1007/978-3-319-72721-9_16

16.2   Numerical Tools for Linear Problems

The numerical tools are an essential part of finite element analysis, and the major numerical tools should be understood by the FE users. Of main interest is the solution of linear equations and of linear eigenvalue problems.   The critical aspects are

- the necessary computer execution time
- the necessary computer storage
- the robustness of the method

This chapter is restricted to problems with symmetric coefficient matrices and describes the Gauss factorization that uses most of the computer execution time for a FE program, say 80%. The superelement technique is an alternative solution technique with several important applications, and this technique is therefore given a special section. The remaining part of the chapter concentrates on numerical tools for solving eigenvalue problems with focus on rate of convergence, shift, and orthogonalization. Finally, the effective method named subspace iteration is described.

16.3   Gauss Factorization

Any non-singular symmetric matrix can be factorized into a product of three matrices, a lower matrix $[l]$, a diagonal matrix $[d]$, and an upper matrix equal to the transposed lower matrix $[l]^T$. For the stiffness matrix $[S]$, this is

$$[S] = [l][d][l]^T \tag{16.3}$$

which is expanded to find more detail formulas for this factorization. An example of a system of order 3 is

$$\begin{bmatrix} S_{11} & S_{12} & S_{13} \\ S_{12} & S_{22} & S_{23} \\ S_{13} & S_{23} & S_{33} \end{bmatrix} = \begin{bmatrix} l_{11} & 0 & 0 \\ l_{21} & l_{22} & 0 \\ l_{31} & l_{32} & l_{33} \end{bmatrix} \begin{bmatrix} d_1 & 0 & 0 \\ 0 & d_2 & 0 \\ 0 & 0 & d_3 \end{bmatrix} \begin{bmatrix} l_{11} & l_{21} & l_{31} \\ 0 & l_{22} & l_{32} \\ 0 & 0 & l_{33} \end{bmatrix}$$
$$= \begin{bmatrix} l_{11}d_1 & 0 & 0 \\ l_{21}d_1 & l_{22}d_2 & 0 \\ l_{31}d_1 & l_{32}d_2 & l_{33}d_3 \end{bmatrix} \begin{bmatrix} l_{11} & l_{12} & l_{13} \\ 0 & l_{22} & l_{23} \\ 0 & 0 & l_{33} \end{bmatrix} \tag{16.4}$$

Note that by post-multiplication with a diagonal matrix (as here $[d]$), each column component is multiplied with the corresponding diagonal element. (With pre-multiplication with a diagonal matrix, each row component is multiplied with the corresponding diagonal element.)

Diagonal components

From the example, (16.4) follows directly that the diagonal components in the matrix $[S]$ are

$$S_{ii} = \sum_{j=1}^{i} l_{ij} d_j l_{ij} = l_{ii}^2 d_i + \sum_{j=1}^{i-1} l_{ij}^2 d_j \tag{16.5}$$

which for Gauss factorization with $l_{ii} = 1$ gives the diagonal components of the diagonal matrix $[d]$

$$d_i = S_{ii} - \sum_{j=1}^{i-1} l_{ij}^2 d_j \tag{16.6}$$

These components are also named the pivot components.

Off-Diagonal components

The off-diagonal components in the matrix $[S]$ are

$$S_{ki} = S_{ik} = \sum_{j=1}^{i} l_{ij} d_j l_{kj} + l_{ii} d_i l_{ki} \tag{16.7}$$

which for Gauss factorization with $l_{ii} = 1$ gives the off-diagonal components of $[l]$

$$l_{ki} = \left(S_{ki} - \sum_{j=1}^{i-1} l_{ij} d_j l_{kj} \right) / d_i \tag{16.8}$$

Bandwidth

A band matrix is a matrix where all components outside a band around the diagonal (diagonal plus co-diagonals) are zero. The bandwidth m for a symmetric matrix is defined as the number of lower co-diagonals plus 1 (the diagonal). This is the integer value important for the necessary computer storage as well as for the necessary computer execution time. The summations in (16.6) and (16.8) can then be limited to the nonzero part.

FORTRAN factorization code

With a chosen storage of "half" the symmetric matrix $[S]$ and a chosen ordering of calculations (16.6) and (16.8), the strategy for computer programming must be chosen. The fact that this calculation takes the major part of the computer execution time for a finite element program makes these few computer lines interesting, and Fig. 16.1 shows the core code of a FORTRAN program, personally used for many years.

Let n denote the number of degrees of freedom and m the bandwidth; a major part of the necessary computer storage is proportional to $n \cdot m$, while a major part (the factorization) of the necessary computer execution time is proportional to $n \cdot m^2$. Note the importance of a moderate bandwidth.

```
*
C       Factorization of symmetric band matrix with
C       diagonal elements in the first row of the matrix
*
        DO 300 N = 1, NUM_DOF - 1
          M = N - 1
          PIV = 1.D0/SYSTEM_FLEXI(1, N)
          DO 150  L = 2, NUM_BANDWIDTH
            IF (SYSTEM_FLEXI(L, N) .EQ. 0.D0) GO TO 150
            I = M + L
            IF (I .GT. NUM_DOF) GO TO 150
            C = SYSTEM_FLEXI(L, N)*PIV
            J = 0
            DO 100 K = L, NUM_BANDWIDTH
              J = J + 1
              SYSTEM_FLEXI(J, I) = SYSTEM_FLEXI(J, I) -
     &                            C*SYSTEM_FLEXI(K, N)
100           CONTINUE
            SYSTEM_FLEXI(L, N) = C
150         CONTINUE
300     CONTINUE
```

Fig. 16.1 FORTRAN code for Gauss factorization of a symmetric band matrix, stored in the array SYSTEM_FLEXI for input as well as for output

16.4 Linear Solutions by Forward and Backward Substitutions

To obtain a linear solution with $[S]$ as coefficient matrix, then in addition to the Gauss factorization, a forward substitution followed by a backward substitution is performed. With multiple load cases (multiple right-hand sides), or with the inverse iterations to be described in Sect. 16.6, the expensive Gauss factorization is needed only once and the substitutions do not take much computer time, compared to the factorization.

Two/three step procedure

Introducing the factorization, the procedure applied to stiffness $[S]$, displacements $\{D\}$, and actions $\{A\}$ follows

$$[S]\{D\} = \{A\} \quad \Rightarrow \quad [L][d][L]^T\{D\} = \{A\}$$

$$\text{solved in a three step procedure:}$$

$$[L]\{\tilde{Y}\} = \{A\} \text{ forward substitution}$$

$$\{Y\} = [d]^{-1}\{\tilde{Y}\} \text{ scalar multiplication}$$

$$[L]^T\{D\} = \{Y\} \text{ backward substitution} \tag{16.9}$$

FORTRAN substitution code

Figure 16.2 shows the core code of a FORTRAN program for this forward and backward substitutions, used in conjunction with the result from the Gauss factorization code.

```
*
C        Forward substitution
*
         DO 200 N = 1, NUM_DOF - 1
            M = N - 1
            C = I_DISPL(N)
            I_DISPL(N) = C/SYSTEM_FLEXI(1, N)
            DO 100 L = 2, NUM_BANDWIDTH
               I = M + L
               I_DISPL(I) = I_DISPL(I) - SYSTEM_FLEXI(L, N)*C
     100    CONTINUE
     200 CONTINUE
*
C        Backward substitution
*
         I_DISPL(NUM_DOF) = I_DISPL(NUM_DOF)/SYSTEM_FLEXI(1, NUM_DOF)
         DO 400 I = 1, NUM_DOF - 1
            N = NUM_DOF - I
            IF (N .EQ. 0) GO TO 400
            M = N - 1
            DO 300 K = 2, NUM_BANDWIDTH
               L = M + K
               I_DISPL(N) = I_DISPL(N) - SYSTEM_FLEXI(K, N)*I_DISPL(L)
     300    CONTINUE
     400 CONTINUE
```

Fig. 16.2 FORTRAN code for forward and backward substitution, with array I_DISPL as input (right-hand side) as well as output (solution)

16.5 Superelement Technique

Solution by a superelement algorithm corresponds to solving the overall finite element equilibrium equation, say of order 10,000–100,000,

$$[S]\{D\} = \{A\} \quad \text{or} \quad \begin{bmatrix} [S]_{cc} & [S]_{cn} \\ [S]_{cn}^T & [S]_{nn} \end{bmatrix} \begin{Bmatrix} \{D\}_c \\ \{D\}_n \end{Bmatrix} = \begin{Bmatrix} \{A\}_c \\ \{A\}_n \end{Bmatrix} \tag{16.10}$$

with static condensation of the non-super degrees of freedom $\{D\}_n$ and thus getting a lower order system, say of order 10 to 100, for the super degrees of freedom $\{D\}_c$. For the case $\{A\}_n = \{0\}$, no condensation of $\{A\}_c$ is involved

$$[S_c]\{D_c\} = \{A_c\} \quad \text{with} \quad [S_c] = [S]_{cc} - [S]_{cn}[S]_{nn}^{-1}[S]_{cn}^T \tag{16.11}$$

where $\{D_c\}$, $\{A_c\}$ are the actual super degrees of freedom and the directly corresponding nodal loads. If $\{A\}_n \neq \{0\}$, the condensed loads are determined by

$$\{A_c\} = \{A\}_c - [S]_{cn}[S]_{nn}^{-1}\{A\}_n \tag{16.12}$$

This superelement technique is well known, and thus, only comments on the practical evaluation of the superelement matrix $[S_c]$ are given. Note that this is just an alternative solution technique and not a further approximation.

Practical evaluation

The practical evaluation of the superelement matrix $[S_c]$ as specified by (16.11) may seem complicated, but from the basic definition of stiffness, as force (at a super dof.) per unit displacement (at the same or another super dof.), follows that a specific column of $[S_c]$, say column k, is determined by a unit forced displacement at the corresponding dof.

Unit forced displacement

All the other contact dof. are fixed to zero, and the calculated reactions corresponding to the super dof. are then the column k of $[S_c]$.

This implies that without performing the matrix calculation in (16.11), the superelement matrix is obtained by solving a number of load cases, corresponding to the number of super dof., and doing so by procedures already available in a finite element code. The computational costs of solving for these additional load cases correspond at most to one additional Gauss factorization, and for larger systems, far less computer execution time is used.

Contact applications

The super finite element technique is recently applied to the solution of contact problems, see Pedersen (2006b, c, 2007).

16.6 Power Method and Inverse Iteration

Methods for solving eigenvalue problems may roughly be classified into

- methods that first find the eigenvalue and then evaluate the eigenvector
- methods that first find the eigenvector and then evaluate the eigenvalue

In short terms, the methods to find eigenvectors are described, and these methods are the most used in relation to FE problems.

Solution separation

Although the power method is not the essential tool for FE eigenvalue problem, it is a natural starting point for this description. For the dynamic system displacement $\{\tilde{D}\}$, assume a separation $\{\tilde{D}\} = \{D\}e^{i\omega t}$ with a dynamic factor $e^{i\omega t}$ where i is the imaginary unit ($i = \sqrt{-1}$), ω is an angular frequency, and t is time, and a time-independent displacement vector $\{D\}$ (amplitudes). Displacements, velocities, and acceleration are then

$$\{\tilde{D}\} = e^{i\omega t}\{D\}, \qquad \frac{d\{\tilde{D}\}}{dt} = i\omega e^{i\omega t}\{D\}, \qquad \frac{d^2\{\tilde{D}\}}{dt^2} = -\omega^2 e^{i\omega t}\{D\} \qquad (16.13)$$

Force equilibrium with system stiffness matrix $[S]$ and system mass matrix $[M]$ without damping and external forces is

$$[M]\frac{d^2\{\tilde{D}\}}{dt^2} + [S]\{\tilde{D}\} = 0 \qquad (16.14)$$

which for the amplitude displacement vector $\{D\}$ gives

$$([S] - \omega^2[M])\{D\} = 0 \quad \text{or}$$

$$[S]\{D\} = \omega^2[M]\{D\} = \omega^2\{Y\} \tag{16.15}$$

where the force vector $\{Y\}$ describes the magnitude of inertia forces, resulting from displacement vector $\{D\}$. Writing (16.15) in standard form

$$[A]\{D\} = \omega^2\{D\} \quad \text{with} \quad [A] = [M]^{-1}[S] \tag{16.16}$$

or on the alternative standard form

$$[A]\{D\} = \frac{1}{\omega^2}\{D\} \quad \text{with} \quad [A] = [S]^{-1}[M] \tag{16.17}$$

Name eigenvector and eigenvalue

Note that the name eigenvector relates to the fact that the vector $\{D\}$ is transformed into itself except for a scalar factor, the eigenvalue ω^2 in (16.16) and $1/\omega^2$ in (16.17).

Name the power method

The power method originates from (16.16), and the iteration procedure is

$$\{X_k\} = [A]\{X_{k-1}\} = [A]^k\{X_0\} \quad \text{with} \quad [A] = [M]^{-1}[S] \tag{16.18}$$

where $\{X_k\}$ is iteration number k based on the initial guess $\{X_0\}$. It is clear that the name power method comes from the iteration factor $[A]$ being raised to the power k.

The method is proved (in Sect. 16.6.1) to converge to the eigenvector $\{D_L\}$ with the largest eigenvalue ω_L^2. Largest is the numerically largest, absolute value.

The inverse iteration method originates from (16.17), and the iteration procedure is

$$\{X_k\} = [A]\{X_{k-1}\} = [A]^k\{X_0\} \quad \text{with} \quad [A] = [S]^{-1}[M] \tag{16.19}$$

where $\{X_k\}$ is iteration number k based on the initial guess $\{X_0\}$.

Inverse iteration

The name inverse iteration method comes from the eigenvalue being an inverse quantity, here $(\omega^2)^{-1}$.

Because the power method is proved to converge to the eigenvector with the largest eigenvalue, the method of inverse iteration will converge to the eigenvector $\{D_S\}$ with the smallest eigenvalue ω_S^2. Smallest is the numerically smallest, absolute value.

16.6.1   Rate of Convergence

To understand the positive as well as the negative aspects of the method of inverse iteration, the proof for convergence is presented. For convenience using $\lambda = \omega^2$ for the eigenvalue. With $\{D_r\}$ being an eigenvector that may be multiplied with any nonzero constant, (16.17) gives

$$[A]\{D_r\} = \frac{1}{\lambda_r}\{D_r\} \quad \text{and} \quad [A]^k\{D_r\} = \frac{1}{\lambda_r^k}\{D_r\} \tag{16.20}$$

where $\{D_r\}$ is the eigenvector corresponding to the eigenvalue λ_r.

Eigenvector space

Any start vector $\{X_0\}$ in inverse iteration (16.19) can be written in the complete space of eigenvectors

$$\{X_0\} = \sum_{r=1}^{n} c_r\{D_r\} \tag{16.21}$$

where n is the order of the problem. This possible expansion is proved for symmetric matrices $[S]$ and $[M]$. Inserting (16.21) in (16.19) and using (16.20) give

$$\{X_k\} = [A]^k\{X_0\} = [A]^k \sum_{r=1}^{n} c_r\{D_r\} = \sum_{r=1}^{n} c_r \frac{1}{\lambda_r^k}\{D_r\} \tag{16.22}$$

which can be rewritten to see the influence of the smallest eigenvalue $\{D_1\}$

$$\{X_k\} = \frac{1}{\lambda_1^k} \sum_{r=1}^{n} c_r (\frac{\lambda_1}{\lambda_r})^k\{D_r\} = \frac{1}{\lambda_1^k}\left(c_1\{D_1\} + \sum_{r=2}^{n} c_r (\frac{\lambda_1}{\lambda_r})^k\{D_r\} \right) \tag{16.23}$$

Proof of convergence

From this last expression, a number of conclusions can be drawn

- Assume $|\lambda_r| > |\lambda_1|$ for all $r > 1$ and $c_1 \neq 0$, then (16.23) proofs $\{X_k\} \Rightarrow \{D_1\}$ for $k \Rightarrow \infty$, except for a factor.
- Assume $|\lambda_r| > |\lambda_2|$ for all $r > 2$ and $c_1 = 0$, then (16.23) proofs $\{X_k\} \Rightarrow \{D_2\}$ for $k \Rightarrow \infty$, except for a factor.
- Assume $|\lambda_2| = |\lambda_1|$ and $|\lambda_r| > |\lambda_2|$ for all $r > 2$ and c_1, c_2 nonzero, then (16.23) shows that $\{X_k\}$ will not converge, but always be a linear combination of $\{D_1\}$ and $\{D_2\}$.
- For $|\lambda_2| \simeq |\lambda_1|$, then (16.23) shows that $\{X_k\}$ will converge slowly.

Proof conclusions

With convergence to $\{D_1\}$, then for a "large" number of iterations k only the influence of $\{D_1\}$ remains

$$\{X_k\} \simeq \frac{c_1}{\lambda_1^k}\{D_r\} \quad \text{and} \quad \{X_{k-1}\} \simeq \frac{c_1}{\lambda_1^{k-1}}\{D_r\} \quad \Rightarrow \quad \frac{\{X_{k-1}\}}{\{X_k\}} \simeq \lambda_1 \qquad (16.24)$$

which should be read in the sense that the eigenvalue is the ratio between components in two following iterations. The fact that this is for any (non zero) part of $\{X_k\}$ means that bounds can be established. Alternatively, the eigenvalue can be evaluated from the Rayleigh quotient.

16.6.2   Orthogonalization and Shift

Energy equilibrium and Rayleigh quotient

The force equilibrium (16.13) is written as an energy equilibrium without the time factor $e^{i\omega t}$, i.e., the potential amplitude equals the kinetic amplitude

$$\{D\}^T[S]\{D\} = \omega^2\{D\}^T[M]\{D\} \qquad (16.25)$$

from which follows the Rayleigh quotient

$$\omega^2 = \frac{\{D\}^T[S]\{D\}}{\{D\}^T[M]\{D\}} \quad \text{or with normalization}$$

$$\omega^2 = \{\Delta\}^T[S]\{\Delta\} \quad \text{when } \{\Delta\}^T[M]\{\Delta\} = 1 \qquad (16.26)$$

i.e., $\{\Delta\} = \{D\}/\sqrt{\{D\}^T[M]\{D\}}$.

Orthogonal properties

With the symmetric problem matrices $[S]$ and $[M]$, the mutual orthogonalization of the eigenvalues is proved

$$\{\Delta_i\}^T[M]\{\Delta_j\} = \delta_{ij} \quad \text{and} \quad \{\Delta_i\}^T[S]\{\Delta_j\} = \delta_{ij}\lambda_i \qquad (16.27)$$

where δ_{ij} is the Kronecker's delta ($i \neq j \Rightarrow \delta_{ij} = 0$ and $i = j \Rightarrow \delta_{ij} = 1$).

The shown orthogonalization property is the tool to avoid convergence to specific eigenvectors, say to $\{\Delta_1\}$, by iterating with vectors for which the corresponding expansion coefficient is zero ($c_1 = 0$). Assuming a non-orthogonalize vector $\{\tilde{X}_k\}$ at iteration k, then subtract the non-wanted part $c_1\{D_1\}$ by making the result orthogonal to $\{\Delta_1\}$

$$({\{\tilde{X}_k\}}^T - c_1\{\Delta_1\}^T)[M]\{\Delta_1\} = 0 \quad \Rightarrow$$

$$\{\tilde{X}_k\}^T[M]\{\Delta_1\} = c_1\{\Delta_1\}^T[M]\{\Delta_1\} = c_1 \tag{16.28}$$

Orthogonalized inverse iteration

A modified inverse iteration procedure with orthogonalization against known pair of eigen pars λ_i, $\{\Delta_i\}$ is then

$$\{\tilde{X}_k\} = [A]\{X_{k-1}\} \quad \text{with} \quad \{X_k\} = \{\tilde{X}_k\} - \sum_{i\ known} c_i\{\Delta_i\}$$

$$\text{with } c_i \text{ known from } c_i = \{\tilde{X}_k\}^T[M]\{\Delta_i\} \tag{16.29}$$

In theory, it should be enough to orthogonalize only in the first iteration, but random computer errors may reintroduce the unwanted eigenvectors. It is therefore suggested to orthogonalize in each iteration step. The shown procedure to find higher order eigenvectors has its limitations and demands high precision with respect to the eigenvectors to be orthogonalized against.

Practical physical procedure

The standard form (16.19) should not be used for practical computation, because $[A] = [S]^{-1}[M]$ may not be a symmetric matrix ($[S]^{-1}$ is symmetric but a product of two symmetric matrices may not be symmetric). Even though a formulation to keep symmetry is possible, then the valuable band structure of the $[S]$ matrix is destroyed. Thus, the practical formulation is the solution like in a statical problem

$$[S]\{\tilde{X}_k\} = [M]\{X_{k-1}\} = \{Y_{k-1}\} \tag{16.30}$$

which also has a more direct physical interpretation as iterating on the displacements resulting from an improved inertia force distribution $\{Y\}$.

Alternative methods or combination methods are necessary. These useful improvements are described next. First, the method using shift of the total eigenvalue spectrum is described. The eigenvalue λ is presented relative to a specified shift value $\bar{\lambda}$, i.e., $\lambda = \bar{\lambda} + \Delta\lambda$.

Spectrum shift

Introducing this shift in (16.15) gives

$$[S]\{D\} = (\bar{\lambda} + \Delta\lambda)[M]\{D\} \quad \Rightarrow \quad ([S] - \bar{\lambda}[M])\{D\} = \Delta\lambda[M]\{D\} \tag{16.31}$$

and thus inverse iteration based on this formulation will converge toward the eigenvector with the smallest distance to the shift (the numerically smallest $\Delta\lambda$). Note that the corresponding eigenvalue can be both below and above the prescribed shift value. The procedure without orthogonalization with reference to (16.30) is

$$([S] - \bar{\lambda}[M])\{\tilde{X}_k\} = [M]\{X_{k-1}\} = \{Y_{k-1}\} \tag{16.32}$$

User controls

The method of shift can be used with or without orthogonalization. The use of shift value $\bar{\lambda}$ includes a number of possibilities for the user,

- control of the searching eigenvalue spectrum (around the shift value)
- control of the convergence rate by influence on the ratios (λ_1/λ_r)

however, the tool to identify what is determined is needed, i.e., the numbers of eigenvalues below the determined eigenvector.

This information is available after the Gauss factorization of $[S] - \bar{\lambda}[M]$, because the number of negative numbers in the diagonal of the factorized results ($[d]$ from (16.6)) equals the number of eigenvalues below the shift value $\bar{\lambda}$. This is often named Sturm sequence check.

16.7 Subspace Iteration

Subspace reality

The method named subspace iteration involves inverse iteration with multiple iteration vectors that are forced to be mutually orthogonal (in the sense of $[M]$ and $[S]$ orthogonal) to avoid converging to equal vectors. This means that iterations are performed with a space spanned by vectors rather than with the vectors themselves. A more detailed description of subspace iteration can be found in Bathe (1996). To relate to the procedures (16.29), (16.32) the subspace iteration in a non-detailed description with shift is

$$([S] - \bar{\lambda}[M])[\tilde{X}_k] = [M][X_{k-1}] \quad \text{with} \quad [X_k] = [\tilde{X}_k][Z] \tag{16.33}$$

where the system matrices ($[S]$, $[M]$) are of order $n \times n$, i.e., of order $10^4 - 10^6$, and the displacement subspace matrix $[X]$ is of order $n \times m$, with m of size 5–20. The matrix $[Z]$ of linear combination factors is thus of order $m \times m$. This matrix is determined after definition of lower order spaces of energies, followed by solution of a new matrix eigenvalue problem, that because of the low order m is solved in full.

Mutual kinetic energies

Define a matrix of mutual kinetic energies $[\tilde{T}]$ of order $m \times m$

$$[\tilde{T}] = [\tilde{X}_k]^T [M][\tilde{X}_k] \tag{16.34}$$

and as the mass matrix $[M]$ is symmetric, so will $[\tilde{T}]$ be. It may be stated that the displacement vectors are projected into a space of mutual kinetic energies.

Mutual potential energies

Also a matrix of mutual potential energies $[\tilde{U}]$ of the same order $m \times m$ is defined

$$[\tilde{U}] = [\tilde{X}_k]^T [S][\tilde{X}_k] \tag{16.35}$$

also being symmetric, thus also projecting the displacement vectors into a space of mutual potential energies.

Complete solution

For the two symmetric matrices, $[\tilde{T}]$ and $[\tilde{U}]$ an eigenvalue problem is defined, either in terms of a single solution λ, $\{Z\}$ or in terms of the full solution $[\lambda]$, $[Z]$ with the diagonal matrix $[\lambda]$ containing all the eigenvalues and the modal matrix $[Z]$ containing column wise all the corresponding eigenvectors.

$$[\tilde{U}]\{Z\} = \lambda[\tilde{T}]\{Z\} \quad \text{or} \quad [\tilde{U}][Z] = [\tilde{T}][Z][\lambda] \tag{16.36}$$

This small-scale problem of order m is solved in full, and thus, there is no problem with close or even multiple eigenvalues. This removes the main critics of the simple inverse iteration and furthermore normally improves the execution time for convergence.

Mutual Orthogonalization

To see that the transformation in (16.33) corresponds to a mutual orthogonalization, the matrix $[T]$ for the transformed subspace $[X_k] = [\tilde{X}_k][Z]$ is

$$[T] = [X_k]^T [M][X_k] = [Z]^T [\tilde{X}_k]^T [M][\tilde{X}_k][Z] = [Z]^T [\tilde{T}][Z] \tag{16.37}$$

where the final expression $[Z]^T [\tilde{T}][Z]$ according to the solved eigenvalue problem (in the energy space) is diagonal. If the matrix $[Z]$ is normalized to give $[Z]^T [\tilde{T}][Z] = [I]$, then the eigenvalues are obtained by $[\lambda] = [Z]^T [\tilde{U}][Z]$. It can be proved that the eigenvalues found are the eigenvalues most close to the shift value $\bar{\lambda}$ in (16.33).

References

Almen, J. O., & Laszlo, A. (1936). The uniform-section disk spring. *Transactions of ASME, 58,* 305–314.

Aluminum,. (1986). *Specifications for aluminum structures.* Washington: American Institute of Steel Construction.

Bathe, K. J. (1996). *Finite Element Procedures* (2nd ed., 1037 pages). New Jersey: Prentice-Hall.

Beck, M. (1952). Die knicklast des einseitig eingespannten, tangential gedruckten stabes. *ZAMP,* 225–228.

Beer, F. P., & Johnston, E. R. (1992). *Mechanics of materials (736 pages).* London: Mc Graw Hill.

Cook, R. D., Malkus, D. S., Plesha, M. E., & Witt, R. J. (2002). *Concepts and applications of finite element analysis* (4th ed., 719 pages). New York: Wiley.

Cowper, G. R. (1966). The shear coefficient in timoshenko's beam theory. *Journal of Applied Mechanics, 33,* 335–340.

Cowper, G. R. (1968). On the accuracy of timoshenko's beam theory. *ASCE, 94*(EM6), 1447–1453.

Crisfield, M. A. (1991 and 1997). *Non-linear finite element analysis of solids and structures* (Vol. 1 and 2, 345 and 494 pages). Chichester: Wiley.

Dzhanelidze, G. Y. (1958). *The stability of a rod acted upon by a slave force.* Leningrad Polytechnic Inst: Trans.

Gantmacher, F. (1959). *The theory of matrices 1.* Chelsea.

Hahn, W. (1967). *Stability of motion.*

Hurwitz, A. (1895). Ueber die bedingungen, unter welchen eine gleichung nur wurzeln mit negativen reellen theilen besitzt. *Mathematische Annalen, 46*(2), 273–284.

Koiter, W. T. (1967). Post-buckling analysis of a simple two-bar frame. In *Recent progress in applied mechanics (The Folke Odqvist volume)* (pp. 337–354).

Koloušek, V. (1973). *Dynamics in engineering structures (translation of Czechoslovakian text) (580 pages).* London: Butterworth.

Lee, G. E., & Reissner, E. (1975). Note on a problem of beam buckling. *Zeitschrift Fur Angewandte Mathematik Und Physik, 26*(6), 839–841.

Liénard, A. M., & Chipart, A. H. (1914). Sur la signe de la partie reelle des racines d'une equation algébrique. *Journal de Mathématiques Pures et Appliquées, 10*(6), 291–346.

Morgan, M. R., & Sinha, S. C. (1983). Influence of a viscoelastic foundation on the stability of Beck's column: An exact analysis. *Journal of Sound and Vibration, 9,* 85–101.

Neer, A., & Baruch, M. (1977). Note on static and dynamic instability of a non-uniform beam. *ZAMP, 28,* 735–740.

© Springer International Publishing AG, part of Springer Nature 2018

S. L. Wiggers and P. Pedersen, *Structural Stability and Vibration,* Springer Tracts in Mechanical Engineering, https://doi.org/10.1007/978-3-319-72721-9

Nikolai, E. L. (1928). *On the stability of the rectilinear form of equilibrium of a bar in compression and torsion*. Leningrad Polytekhn: Inv.

Panovko, Y. G., & Gubanova, I. I. (1964). *Stability and oscillations of elastic systems* (290 pages). Consultants Bureau.

Pedersen, N. L., & Pedersen, P. (2011). Stiffness and design for strength of trapezoidal Belleville springs. *The Journal of Strain Analysis for Engineering Design, 46*(8), 825–836.

Pedersen, P. (1973). Optimal joint positions for space trusses. *Journal of the Structural Division, ASCE, 99*(ST 12), 2459–2476.

Pedersen, P. (1977a). Influence of boundary conditions on the stability of a column under non-conservative load. *International Journal of Solids and Structures, 13*, 445–455.

Pedersen, P. (1977b). On computer-aided analytical element analysis and the similarities of tetrahedron elements. *International Journal for Numerical Methods in Engineering, 11*, 611–622.

Pedersen, P. (1986). A note on vibration of beam-columns. *Journal of Sound and Vibration, 105*(1), 143–150.

Pedersen, P. (2005a). Analytical stiffness matrices with Green-Lagrange strain measure. *International Journal for Numericals Methods in Engineering, 62*, 334–352.

Pedersen, P. (2005b). Axisymmetric analytical stiffness matrices with Green-Lagrange strains. *Computational Mechanics, 35*, 227–235.

Pedersen, P. (2006a). Analytical stiffness matrices for tetrahedral elements. *Computer Methods in Applied Mechanics and Engineering, 196*, 261–278.

Pedersen, P. (2006b). A direct analysis of elastic contact. *Computational Mechanics, 37*(1), 221–231.

Pedersen, P. (2006c). On shrink fit analysis and design. *Computational Mechanics, 37*(2), 121–130.

Pedersen, P. (2007). On the influence of clearance in orthotropic disc-pin contacts. *Composite Structures, 79*, 554–561.

Popov, E. (1962). *The dynamics of automatic control systems*. Pergamon Press.

Rao, S. S. (2007). *Vibration of continuous systems (720 pages)*. New Jersey: Wiley.

Roorda, J. (1965). Stability of structures with small imperfections. *ASCE, 91*(EM1), 87–106.

Roorda, J., & Chilver, A. H. (1970). Frame-buckling, an illustration of the perturbation technique. *International Journal of Non-Linear Mechanics, 5*, 235–246.

Routh, E. J. (1955). *Advanced rigid dynamics (from 1875)* (p. 210). New York: Dover.

Shabana, A. A. (1997). *Vibration of discrete and continuous systems* (2nd ed., 393 pages). New York: Springer.

Steel,. (1989). *Manual of steel constructions* (9th ed.). New York: American Institute of Steel Construction.

Thomsen, J. J. (2003). *Vibrations and stability, advanced theory, analysis and tools* (2nd ed., 404 pages). Berlin: Springer.

Timber,. (1985). *Timber construction manual*. New York: Wiley.

Timoshenko, S. P., & Gere, J. M. (1961). *Theory of elastic stability* (2nd ed.). New York: McGraw-Hill.

Volterra, E., & Zachmanoglou, E. C. (1965). *Dynamics of vibration*. Merrill Books.

Wittrick, W. H. (1982). On a differential equation occurinng in structural mechanics: A simple aid to computing. *International Journal of Numerical Methods in Engineering, 18*, 1733–1735.

Wolfram, S. (1991). *Mathematica: A system for doing mathematics by computer*. Addison-Wesley.

Ziegler, H. (1968). *Principles of structural stability* (150 pages). Blaisdell.

Index

A
Asymptotically stable, 134

B
Backward substitution, 144
Bandwidth, 143
Basic assumptions, 6
Basic Berry function, 27
Belleville springs, 114
Bernoulli- Euler beam, 5
Berry functions, 3, 25
Boundary conditions, 10
Buckling force, 49
Build up column, 109

C
Chaos theory, 1
Characteristic equation, 116
Charnier, 112
Code of Fortran program, 143
Co diagonals, 143
Combination of beam-columns, 85
Complete solution, 152
Complete space of eigenvectors, 148
Complex conjugated roots, 138
Computer execution time, 142
Computer storage, 142
Concept of stability, 3
Condensation of $\{A\}$, 145
Condensation of $[S]$, 145
Conjugated, 125
Contact applications, 146
Continuously supported, 53
Control of the convergence rate, 151
Control of the searching spectrum, 151

Cookbook, 89
Correct dimensions, 8
Critical damping, 120

D
d'Alembert principle, 8, 80
Damped single dof vibrations, 121
Damping ratio, 122
Degenerated model, 80
Design formulas, 101
Design of coned disk springs, 114
Determinant condition, 116
Diagonal components, 143
Diagonal matrices, 142
Different eigenvalue problems, 9
Different stable position, 113
Discretized models, 4
Displacement gradient stiffness matrix, 124
Divergence instability, 4
Divergent instability, 49, 83
Dynamic instability, 49

E
Eccentric column loading, 98
Eigenfrequency modes, 17
Eigenvalue $\Rightarrow$ eigenvector, 146
Eigenvalue evaluation, 149
Eigenvector $\Rightarrow$ eigenvalue, 146
Eigenvector space, 148
Elastic snap through, 112
Elliptical integral, 107
Energy equilibrium, 149
Energy method, 48
Engesser in 1889, 99
Equivalent length, 95

© Springer International Publishing AG, part of Springer Nature 2018
S. L. Wiggers and P. Pedersen, *Structural Stability and Vibration*, Springer Tracts
in Mechanical Engineering, https://doi.org/10.1007/978-3-319-72721-9

Euler BC, **13**
Euler case I, 14, 17
Euler case II - IV, 15
Euler case II - VI, 18
Euler case V, 16
Euler cases, 3, 13
Even Hurwitz determinants, 136
Explicit procedure, 131
Explicit solutions, 63
External damping, 71

F
Factorization, 142
Factor of safety, 100
Fail safe, 111
Flexibilities, 75
Flutter instability, 4, 71
Follower column force, 70
Fortran factorization code, 143
Fortran substitution code, 144
Forward substitution, 144
Frequency functions, 45
Frequency parameter, 78

G
Gauss factorization, 142
Generalized Berry functions, 44
General solution, 60
Green-Lagrange strain, 124

H
Harmonic vibrations, 79, 120
Higher order eigenvectors, 150
High-frequency effects, 1
Hurwitz determinants, 135
Hurwitz in 1895, 134

I
Imperfection method, 48
Imperfection sensitive, 4, 91
Implicit procedure, 129
Implies instability, 116
Improved beam modeling, 102
Incremental, iterative solution, 129
Inertia force distribution, 150
Inertia forces, 147
Instability formulations, 48
Instability modes, 14
Integrated treatment, 2
Integration of Berry function, 28

Inverse approach, 63
Inverse iteration, 146, 147
Inverse problem, 50
Inverse solution, 107

K
Karman in 1910, 99
Kinetic energy space, 151
Kirchhoff's dynamic analogy, 107

L
Large displacements, 105
Large strains, 105
Largest, absolute eigenvalue, 147
L'Hospital's rule, 25, 28
Liénard–Chipart in 1914, 134
Limiting cases, 46
Linear bifurcation buckling, 128
Linear combination, 119
Linear combination factors, 151
Linear eigenvalue problems, 142
Linear equations, 142
Linear, free vibrations, 133
Local and global buckling, 111
Lower matrices, 142

M
Magnetic attraction, 57
Matching the BC, 26
Matrix notation, xiii, 2
Method for a scalar function, 141
Method for a vector function, 141
Mikhailov geometrical criteria, 138
Modal analysis, 119
Modeling conclusions, 104
Models, 5
Multiple iteration vectors, 151
Multiple load cases, 144
Mutual elastic energy, 117
Mutual kinetic energies, 151
Mutual kinetic energy, 117
Mutual orthogonalization, 149, 152
Mutual potential energies, 152

N
Name eigenvector eigenvalue, 147
Name power method, 147
Names and symbols, 10
Necessary and sufficient criteria, 134, 136
Necessary criterion, 134

Newton-Raphson iterations, 141
Newton-Raphson method, 45
Non singular matrix, 142
Non-damped free vibration, 119
Non-damped single dof vibrations, 120
Non-dimensional, 7
Non-dimensional frequency, 9
Non-dimensional spring, 36
Non-dimensional spring stiffnesses, 29
Non-linear beam-column, 24
Non-trivial solution, 14
Notations, 2
Number of degrees of freedom, 115
Numerical tools, 142

O
Odd Hurwitz determinants, 136
Off diagonal components, 143
Orthogonal eigenmodes, 118
Orthogonalization, 149
Orthogonalize inverse iteration, 150
Orthogonal properties, 149
Overview of the results, 21

P
Partial follower force, 82
Pivot components, 143
Polynomial problem, 134
Polynomials with complex coefficients, 138
Positive definite matrix, 118
Post-critical imperfection analysis, 91
Post-multiplication, 142
Potential energy space, 152
Power method, 146
Practical evaluation, 145
Practical physical procedure, 150
Pre-multiplication, 142
Proof conclusions, 148
Proof of convergence, 148

Q
Quasi-static, 49

R
Rate of convergence, 148
Rayleigh damping, 119
Rayleigh quotient, 149
Real, positive eigenfrequencies, 117
Residuals and Newton–Raphson iterations,
126

Robustness, 142
Routh in 1875, 134
Routh–Hurwitz criteria, 133
Routh–Hurwitz-Liénard–Chipart   criteria,
133

S
Secant constitutive matrix, 125
Secant formula, 98
Secant relations, 125
Secant stiffness matrix, 124
Self-adjoint, 118
Separation, 133
Seven defined sub-functions, 60
Shanley in 1946, 100
Shift, 149
Shift value, 150
Simple beam theory, 5, 24
Simplifications by Liénard–Chipart, 133
Simply supported, 9
Slenderness ratio, 95
Smallest, absolute eigenvalue, 147
Snap through mechanism's, 114
Solution procedure, 89
Solution separation, 146
Spectrum shift, 150
Stability parameter, 78
Stable and attractive, 134
Stable vibrations, 121
Standard form, 147
Static condensation, 145
Stiffness matrix depends, 124
Stiffnesses, 75
Strain/displacement vector, 124
Stress stiffness matrix, 124
Sturm sequence check, 151
Sub-determinant, 135, 136
Subspace iteration, 151
Subspace iteration method, 115
Subspace reality, 151
Super degrees of freedom, 145
Superelement technique, 145
System mass matrix, 146
System stiffness matrix, 146

T
Tangent constitutive relation, 126
Tangent relations, 125
Tangent stiffness matrix, 124
Tetrahedron element, 127
The Elastica, 105
Timoshenko, 95, 104

Traditional // notations, xiii
Transcendental equations, 62
Translational spring, 36
Trivial solution, 49
Truss column, 109
Two cantilever cases, 38
Two/three step procedure, 144

U
Uniform beam, 7
Unit forced displacement, 146
Unstable vibrations, 121

Upper matrices, 142
User controls, 151

V
Vibrating pendulum, 107
Virtual work, 125
Volume force, 8
Von Mises truss, 112

W
Winkler support, 3, 53

The manufacturer's authorised representative in the EU is Springer
Nature Customer Service Centre GmbH, Europaplatz 3, 69115 Heidelberg,
Germany. If you have any concerns regarding our products, please
contact ProductSafety@springernature.com

Printed and bound by CPI Group (UK) Ltd, Croydon, CR0 4YY
28/11/2025
02007732-0007